CONTENTS

1 Introduction

Who is this book for?

What is the purpose of this book?

What is in this book?

How should I use this book?

WHO IS THIS BOOK FOR?

This book has been written to help anyone who is studying for the Higher Physics examination.

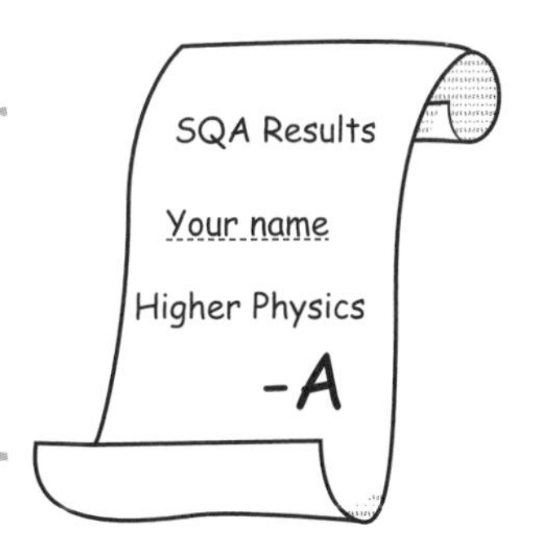

WHAT IS THE PURPOSE OF THIS BOOK?

This book is written specifically to help students gain the best grade they can achieve in Higher Physics. It has been designed to help you to:

- avoid common pitfalls which cause students to lose marks in the final examination
- develop your ability to give answers of a better quality in the final examination

WHAT IS IN THIS BOOK?

This book clarifies the structure and contents of the Higher Physics course and the assessment leading to a Course Award.

This book lists and clarifies reasons why many students fail to gain better marks in the final examination than they could achieve.

This book contains advice and examples on how to avoid losing marks in examination questions.

It also contains advice and examples of how to gain even more marks in many types of examination question.

There are sample questions and example answers to improve your abilities both to identify good answers and to improve the quality of your own answers.

HOW SHOULD YOU USE THIS BOOK?

Use this book alongside your course notes and/or textbook.

Use it regularly so you can train yourself how to transfer the knowledge from your notes into very good (or excellent) answers – so gaining more marks in the final examination.

It will be particularly useful during the final run-up to the examination when you need as much practice as possible at answering past paper questions and reviewing the quality of your own answers.

2 Higher Physics – Course Content

The units of the Higher Physics Course

What is covered in each unit?

THE UNITS

There are three units in the Higher Physics Course. These are:

1. Mechanics and Properties of Matter
2. Electricity and Electronics
3. Radiation and Matter

WHAT IS COVERED IN EACH UNIT?

A brief breakdown of the content covered in each unit is given below.

1. Mechanics and Properties of Matter
   - Vectors
   - Equations of Motion
   - Newton's Second Law, Energy and Power
   - Momentum and Impulse
   - Density and Pressure
   - Gas Laws
2. Electricity and Electronics
   - Electric Fields and Resistors in Circuits
   - Alternating Current and Voltage
   - Capacitance
   - Analogue Electronics

3. Radiation and Matter
   - Waves
   - Refraction of Light
   - Optoelectronics and Semiconductors
   - Nuclear Reactions
   - Dosimetry and Safety

It is recommended that you have to hand the *Content Statements* for each of these units and sub-units. The statements are a detailed list of the things you should know and be able to do for the final examination. They form the basis of the content and structure of the questions when the examination is being written. Complete familiarity with the content statements is therefore essential for top success in the examination.

These lists are contained in the Arrangements document for the course and can be downloaded from the SQA website (www.sqa.org.uk/ then choose Physics).

3 Assessment and the Examination

Unit tests

Practical assessment

The National Examination

THE UNIT TESTS

Your final course award for Higher Physics depends on your performance in the final examination, not on how high a mark you score in the unit tests. However, you must *pass* one test for each of the units to be eligible for a course award from the SQA. A unit assessment (or test) examines both Knowledge and Understanding and Problem Solving.

Each assessment lasts for 45 minutes and consists of a total of 30 marks. The threshold of attainment (the pass mark) is 18 marks (i.e. 60% of the total). More than one attempt to pass a unit assessment is permitted.

PRACTICAL ASSESSMENT

During the course, you will have been actively involved in at least one experiment and have completed a write-up to certain standards laid down by the SQA.

If necessary, you should be given guidance about the required standards and an opportunity to redraft the write-up if it did not meet those standards initially. *You must successfully complete one experimental write-up* to be eligible for a course award from the SQA.

THE NATIONAL EXAMINATION

The examination consists of one paper lasting 2½ hours. There are two sections in the paper:

- Section A has 20 marks for 20 multiple choice questions.
- Section B has 70 marks for extended answer questions.

All questions should be attempted.

The marks are evenly divided across the units in the course – i.e. there are about 30 marks for each of the three units.

Approximately 40% of the total 90 marks are classified as Knowledge and Understanding and 60% Problem Solving. The differences between Knowledge and Understanding and Problem Solving are explained in chapters 4 and 5.

To achieve an award at 'A' or 'B', you need to gain most of the marks for Knowledge and Understanding *and* a high proportion of the Problem Solving marks.

4 Knowledge and Understanding

What is Knowledge and Understanding?

How good does my Knowledge and Understanding have to be?

How can I improve my Knowledge and Understanding?

WHAT IS KNOWLEDGE AND UNDERSTANDING?

'Knowledge and Understanding' questions are about being able to recall facts, symbols, diagrams, ideas and techniques. Being able to use relationships (formulas) to carry out straightforward calculations is also Knowledge and Understanding.

For example, the following questions are past examination questions designed to test knowledge and understanding.

EXAM EXAMPLE 1

Which of the following contains one scalar quantit
and one vector quantity?

A acceleration; displacement

B kinetic energy; speed

C momentum; velocity

D potential energy; work

E power; weight

To answer this question you need to:

- *know* the terms 'scalar' and 'vector'
- *have learned* which quantities are scalars and which are vectors

The answer is E.

EXAM EXAMPLE 2

Which of the following proves that light is transmitted as waves?

A Light has a high velocity.

B Light can be reflected.

C Light irradiance reduces with distance.

D Light can be refracted.

E Light can produce interference patterns

To answer this question you need to:

- *know* that the proof of wave motion is that an interference pattern can be produced, for example by learning content statement 3.1.6 (State that interference is the test for a wave.)

The answer is E.

EXAM EXAMPLE 3

A cylinder of compressed oxygen gas is in a laboratory.

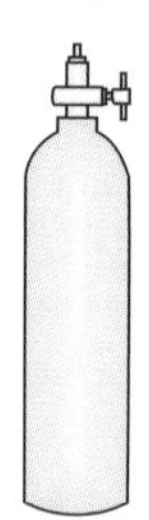

(*a*) The oxygen inside the cylinder is at a pressure of $2{\cdot}82 \times 10^{6}$ Pa and a temperature of 19·0 °C.

The cylinder is now moved to a storage room where the temperature is 5·0 °C.

(i) Calculate the pressure of the oxygen inside the cylinder when its temperature is 5·0 °C.

To answer this question you need to:

- *know* how to carry out a calculation using the gas law
- *remember* to change temperatures into Kelvin values

The answer is:

$$\frac{P_1}{T_1} = \frac{P_2}{T_2}$$

$$\frac{2{\cdot}82 \times 10^6}{292} = \frac{P_2}{278}$$

$$\Rightarrow P_2 = 2{\cdot}68 \times 10^6 \text{ Pa}$$

EXAM EXAMPLE 4

Using the terms *electrons, holes* and *photons*, explain how light is produced at the p-n junction of the LED.

To answer this question you need to have:

- *learned* to describe the production of light in an LED, for example from content statement 3.3.29. (State that in the junction region of a forward-biased p–n junction diode, positive and negative charge carriers may recombine to give quanta of radiation.)

The answer is 'electrons and holes recombine at the junction and produce photons of light'.

EXAMPLE 5

What is meant by an *activity of* 8 kBq?

To answer this question you need to have:

- *learned* the definition of 'activity', for example from content statement 3.5.1. (State that the activity of a radioactive source is the number of decays per second and is measured in becquerels (Bq), where one becquerel is one decay per second.)

The answer is '8 000 nuclei are decaying each second'.

HOW GOOD DOES MY KNOWLEDGE AND UNDERSTANDING HAVE TO BE?

The simple answer to this question is 'It has to be very good'.

Obviously, in order to gain a top grade in Higher Physics you have to score a high mark. The problem-solving marks are generally harder to achieve, so you cannot afford to lose many of the marks for Knowledge and Understanding.

HOW CAN I IMPROVE MY KNOWLEDGE AND UNDERSTANDING?

You must:

- ensure you know your notes/textbook very well – this means re-reading them regularly and then trying to write out facts, symbols etc. from memory
- practise answering questions on a regular basis to ensure you are familiar with all the formulas in the course
- get to know the Content Statements well

Appendix A (page 141) contains a bank of 10 questions designed to improve your ability to use the formulas met during the Higher Physics course. Worked answers to these questions are also included.

For loads more practice calculation questions and answers, go to www.leckieandleckie.co.uk/7248calc.pdf

5 Problem Solving

What is Problem Solving?

How good does my Problem Solving have to be?

How can I improve my Problem Solving?

WHAT IS PROBLEM SOLVING?

'Problem Solving' is about being able to apply your knowledge and understanding so that you can answer questions which are presented in a way or context slightly different to what you are likely to have met.

Approximately half of the Problem Solving marks in the final examination are for questions of a 'more complex nature' which may be set in a more complicated context or be presented in a less structured way.

The following questions are past examination questions designed to test 'straightforward' problem solving.

EXAM EXAMPLE 1

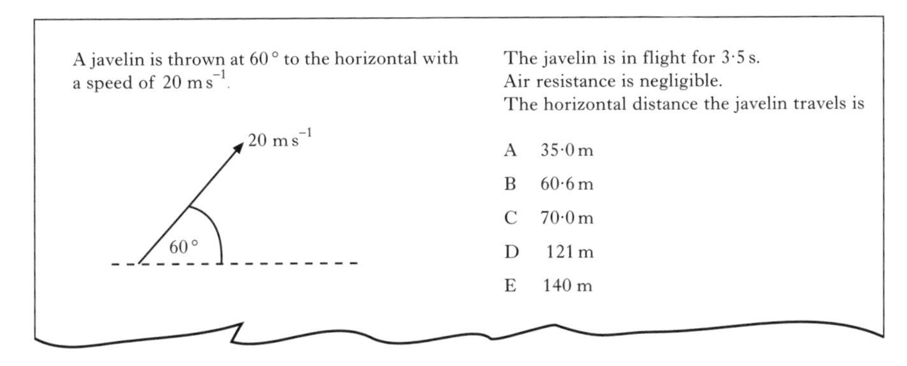

A javelin is thrown at 60 ° to the horizontal with a speed of 20 m s^{-1}.

The javelin is in flight for 3·5 s.
Air resistance is negligible.
The horizontal distance the javelin travels is

A 35·0 m

B 60·6 m

C 70·0 m

D 121 m

E 140 m

To answer this question you need to:

- *be able to find* the horizontal component of the velocity vector (no formula is listed in the Physics data booklet)
- *know* that the horizontal component remains constant
- *apply* the distance, speed, time relationship to horizontal velocity

The answer is:

Horizontal component of velocity = $v \cos \theta = 20 \cos 60 = 10\ \text{m s}^{-1}$

Horizontal distance = speed × time = $10 \times 3{\cdot}5 = 35$ m

The answer is A.

EXAM EXAMPLE 2

A person stands on a weighing machine in a lift. When the lift is at rest, the reading on the machine is 700 N. The lift now descends and its speed increases at a constant rate. The reading on the machine

A is a constant value higher than 700 N

B is a constant value lower than 700 N

C continually increases from 700 N

D continually decreases from 700 N

E remains constant at 700 N.

To answer this question you need to:

- *know* that an object at rest has balanced forces acting on it
- *deduce* that the weight of the object is 700 N
- *know* that an object which is accelerating uniformly has a *constant* unbalanced force on it
- *apply* Newton's second law to realise that the unbalanced force is, in this case, downwards
- *deduce* that the reading is constant and must be less than the weight

The answer is B.

EXAM EXAMPLE 3

A metal plate emits electrons when certain wavelengths of electromagnetic radiation are incident on it.

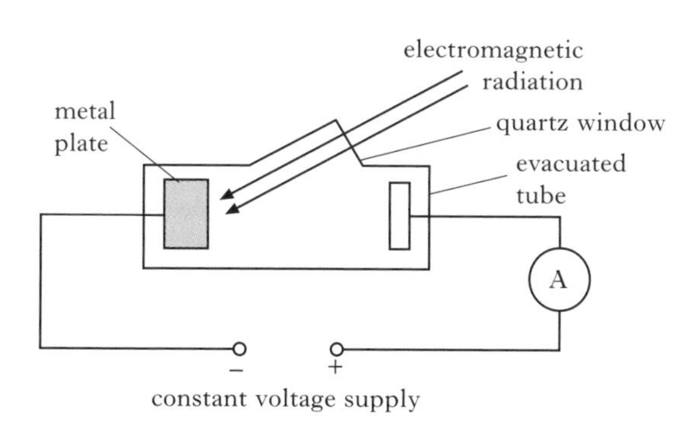

When light of wavelength 605 nm is incident on the metal plate, electrons are released with zero kinetic energy.

(*a*) Show that the work function of this metal is $3 \cdot 29 \times 10^{-19}$ J.

To answer this question you need to put together various bits of what you know and apply them as follows:

- *know* that 'work function' means the minimum energy an electron needs to be able to escape from the surface of a metal
- *know* that the energy to escape is given to electrons by photons in the incident radiation
- *know* that one photon gives all its energy to one electron
- *know and apply* the formula to calculate the energy of a photon ($E = hf$)
- *know* that frequency, f, can be calculated from the wave formula (i.e. $f = \frac{v}{\lambda}$)
- *realise* that another relationship which applies is $E_k = hf - hf_o$ and that, in this case, work function $= hf_o = hf$ (because the kinetic energy is zero)

The answer is:

$$\text{work function} = hf_o = hf = h\frac{v}{\lambda} = 6{\cdot}63 \times 10^{-34} \times \frac{3{\cdot}0 \times 10^{8}}{605 \times 10^{-9}}$$
$$= 3{\cdot}28760 \times 10^{-19}$$
$$= 3{\cdot}29 \times 10^{-19}\ \text{J}$$

For full marks, you also need to remember that the prefix 'nano' ('n') means '10^{-9}'.

The following is an example from the 2008 examination of a question which is designed to test 'more complex' problem solving.

EXAM EXAMPLE 4

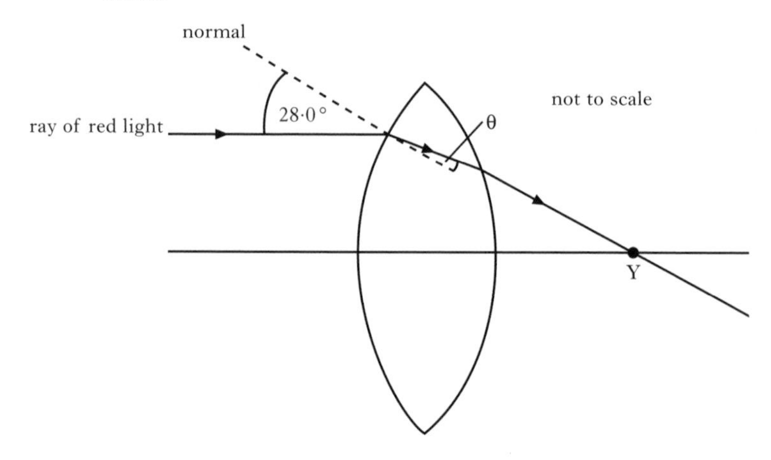

(*a*) A ray of red light of frequency $4{\cdot}80 \times 10^{14}$ Hz is incident on a glass lens as shown.

The ray passes through point Y after leaving the lens.

The refractive index of the glass is 1·61 for this red light.

(ii) Calculate the wavelength of this light inside the lens.

To answer this question you need to:

- *know* that there is no change in frequency when the light refracts into the glass
- *know* that speed and wavelength *do* change when the light refracts into the glass

- *know* (or look up) the speed of light in air = $3{\cdot}0 \times 10^8$ m s^{-1}
- *select and combine* two relationships listed in the Physics data booklet, namely

  $$n = \frac{\sin\theta_1}{\sin\theta_2} \quad \text{and} \quad \frac{\sin\theta_1}{\sin\theta_2} = \frac{\lambda_1}{\lambda_2} = \frac{v_1}{v_2}$$

- *realise* that another relevant relationship is the wave formula, $v = f\lambda$
- *apply* the wave formula to calculate either the speed or the wavelength in air
- *apply* a combination of the other two relationships to calculate the wavelength in the glass
- *present* the analysis in a clear, complete and logical way

An exemplar answer would be:

In air, $\lambda = \frac{v}{f} = \frac{3{\cdot}0 \times 10^8}{4{\cdot}8 \times 10^{14}} = 6{\cdot}25 \times 10^{-7}$ (m)

On refraction, $\frac{\lambda_1}{\lambda_2} = \frac{\sin\theta_1}{\sin\theta_2} = n$

$$\text{so } \lambda_2 = \frac{\lambda_1}{n}$$
$$= \frac{6{\cdot}25 \times 10^{-7}}{1{\cdot}61}$$
$$= 3{\cdot}881988 \times 10^{-7}$$
$$= 3{\cdot}9 \times 10^{-7}\ \text{m}$$

HOW GOOD DOES MY PROBLEM SOLVING HAVE TO BE?

The simple answer to this question is 'It has to be good or very good'.

To achieve an award at 'A' or 'B' you must gain most of the marks for Knowledge and Understanding *and* a high proportion of the marks for Problem Solving.

HOW CAN I IMPROVE MY PROBLEM SOLVING?

You must:

- ensure you have an excellent knowledge base – very few problem solving questions can be attempted successfully if you have not learned the facts listed in the Content Statements
- practise answering as many past paper questions as possible, especially those which require you to 'describe', 'explain' or 'justify'.

Appendix B (page 143) contains a bank of questions designed to improve your ability to choose the appropriate Physics to explain a variety of different situations. Answers to these questions are also included.

For loads more practice at problem solving questions and answers, go to www.leckieandleckie.co.uk/7248prob.pdf

6 Learning from the mistakes of previous candidates (Advice from the SQA)

Every year the SQA publishes a report on the performance of candidates in each national examination. Amongst other things, these reports identify reasons why marks are lost and give advice on how candidates could be better prepared for the examination in the future.

These External Assessment Reports on Higher Physics are therefore an essential tool for anyone wanting to succeed at the top levels.

The issues raised in these reports have been used in constructing the following reasons why candidates have lost marks in previous examinations. The majority of the rest of this book concentrates on techniques to help you perform well in these areas.

ADDRESSING AREAS OF WEAKNESS IN MECHANICS AND PROPERTIES OF MATTER

Vectors

- **Failing to draw arrows on lines in vector diagrams**

An important fact about any vector quantity is that it has a direction. The line representing the quantity in a vector diagram needs to have its direction shown by drawing an arrow on the line. This will also help to ensure that you correctly draw the vectors 'nose to tail' in order to add them together.

EXAM EXAMPLE 1

Competitorsare racing remote control cars. The cars have to be driven over a precise route between checkpoints.

Each car is to travel from checkpoint A to checkpoint B by following these instructions.

"Drive 150 m due North, then drive 250 m on a bearing of 60° East of Nortl (060)."

(*a*) By scale drawing or otherwise, find the displacement of checkpoint B fro checkpoint A.

Answer:

60°

38°

Displacement = 350m at 038 / 38° E of N

It is important to remember to draw the resultant from the start point to the end point to make sure its direction is correct.

Arrows should be shown on all vectors.

- **Vector diagrams are drawn too small and inaccurately**

The accuracy of your answer depends on how precisely you have drawn the lengths of the lines and the angles between them. It is generally true that the larger your diagram, the more accurate the measurements although, of course, there is a limit to its maximum size. As an approximate guide, choose a scale to make your diagram about one third to one half of an A4 page.

Remember that three figure bearings are always with respect to North and you may have to include reference lines in your diagram – like the dashed line in the above example.

- **Vectors not added 'nose-to-tail'**

Many candidates make the mistake of drawing both (or all) of the vectors from the same point and then try to find the resultant by joining the ends of the vectors – this is *always* wrong.

You must choose a suitable starting point (origin), making allowance for how your drawing will go up/down and left/right across the page. The first vector is then drawn from that point to its correct size (according to the scale you have chosen to use) and in its correct direction. The next vector is then drawn *from the end point of the first vector*. Finally, once all the vectors have been drawn, *the resultant is found by drawing a straight line from the origin to the end of the last vector*.

For example:

EXAM EXAMPLE 2

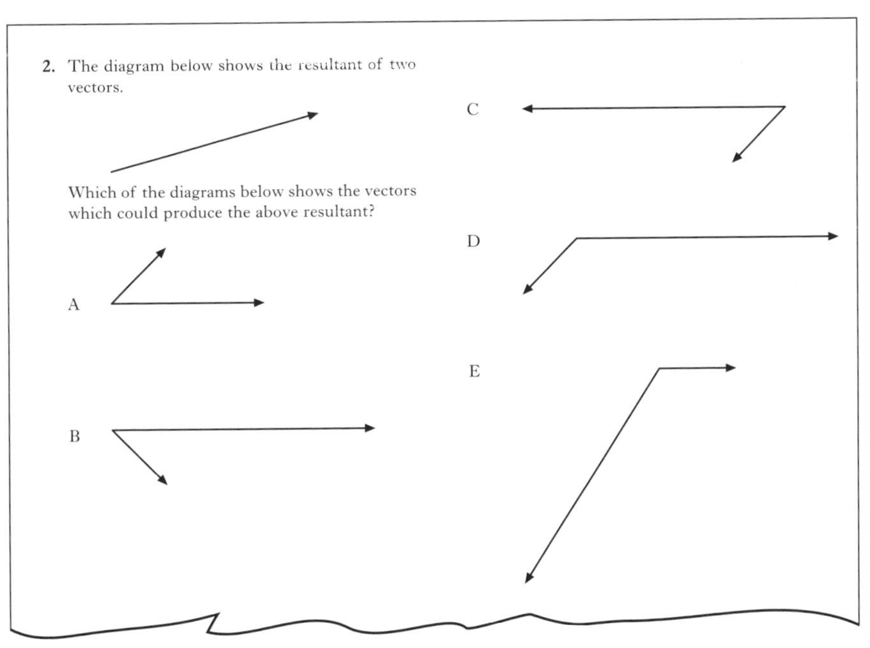

2. The diagram below shows the resultant of two vectors.

Which of the diagrams below shows the vectors which could produce the above resultant?

A

B

C

D

E

Option A is the correct answer.

The diagrams show vectors before they are added.

In order to add them together, one of them needs to be drawn at the end of the other one. Therefore, for the vectors in option A:

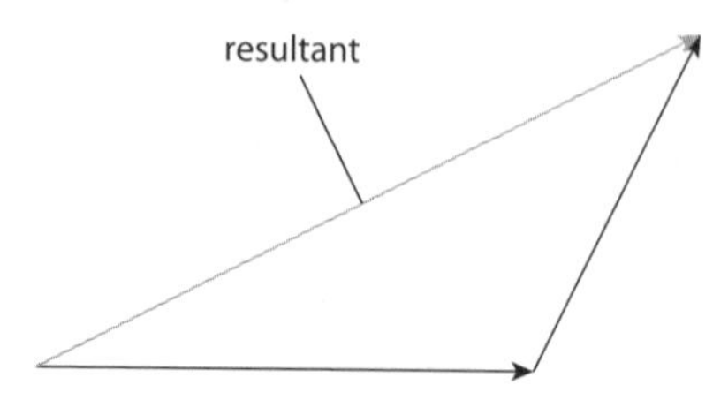

It is also important to note that the order of adding the vectors does not affect the answer – the same resultant is produced:

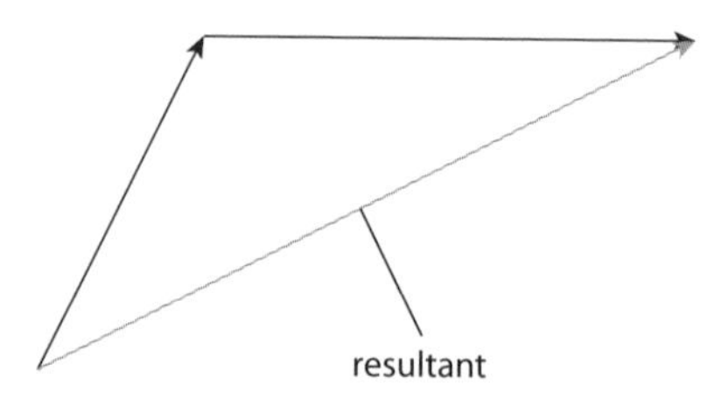

- **Failing to quote a direction for the resultant vector**

A vector quantity is defined as one which requires a direction as well as a value (or magnitude). When you are asked to find a resultant displacement, a resultant velocity or a resultant force, your answer is not complete unless you have included the *direction* of that resultant.

- **Using Pythagoras' formula in triangles which are not right-angled**

Pythagoras' theorem states that the square on the hypotenuse, c, is equal to the sum of the squares on the other two sides (a and b), i.e. $c^2 = a^2 + b^2$

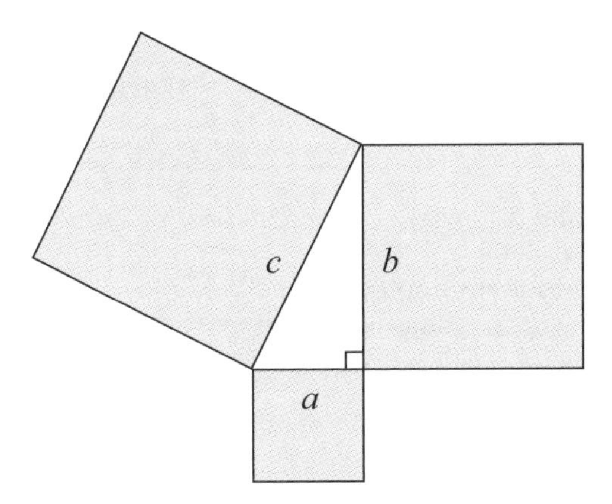

The Pythagoras rule is *only* applicable to *right angled* triangles.

Some questions require vector additions which produce triangles which are not right angled. In these cases you must use a scaled vector diagram to find the resultant. Or, if you wish, you may use the more advanced mathematics of the cosine rule and the sine rule, i.e.

$$a^2 = b^2 + c^2 - 2bc\cos A$$

$$\frac{a}{\sin A} = \frac{b}{\sin B} = \frac{c}{\sin C}$$

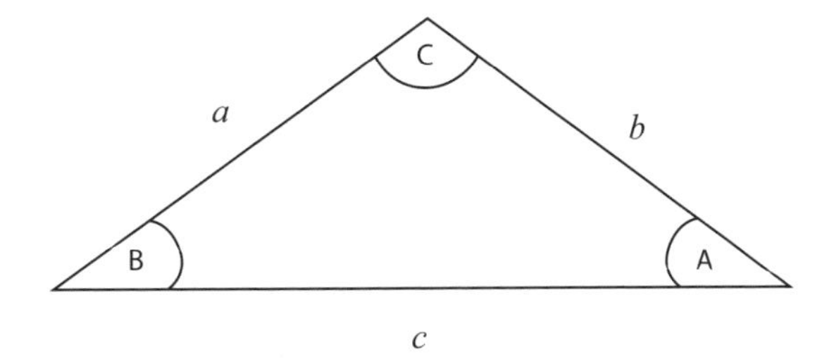

- **Lack of understanding of the difference between 'distance' and 'displacement'**

'Distance' is a scalar. The total distance is the sum of the parts of the journey.

'Displacement' is a vector. The resultant displacement is the straight line from the start to the finish. The resultant displacement of an object usually has a smaller value than the total distance travelled.

Question 1 in the 2005 SQA paper is a good example for both this issue and the following one.

- **Lack of understanding of the difference between 'average speed' and 'average velocity'**

$$\text{Average speed} = \frac{\text{total distance}}{\text{time}}$$

$$\text{Average velocity} = \frac{\text{resultant displacement}}{\text{time}}$$

The resultant velocity usually has a smaller value than the average speed.

EXAM EXAMPLE 3

A car travels from X to Y and then from Y to Z as shown.

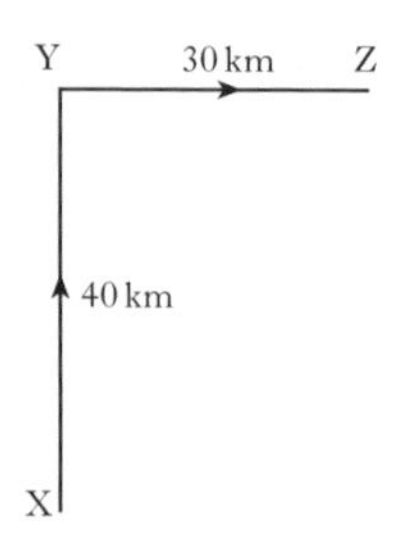

It takes one hour to travel from X to Y. It also takes one hour to travel from Y to Z.

Which row in the following table shows the magnitudes of the displacement, average speed and average velocity for the complete journey?

	Displacement (km)	*Average speed* (km h^{-1})	*Average velocity* (km h^{-1})
A	50	35	25
B	70	35	25
C	50	35	35
D	70	70	50
E	50	70	25

The total distance travelled is 40 + 30 = 70 km.
The value of the resultant displacement is the length of the line from X to Z. This is a right angled triangle (although the question does not state this!), so Pythagoras can be used to calculate the length of the line XZ as 50 km.

$$\text{Average speed} = \frac{\text{total distance}}{\text{time}} = \frac{70\text{ km}}{2\text{ h}} = 35\text{ km h}^{-1}$$

$$\text{The value of the average velocity} = \frac{\text{total displacement}}{\text{time}} = \frac{50\text{ km}}{2\text{ h}} = 25\text{ km h}^{-1}$$

The correct answer is A.

(Note that this question only tests your understanding about *values* of speed and velocity. It is not testing your understanding that velocity and displacement also need a direction in order to be fully defined.)

- **Lack of knowledge of the formulas for the rectangular components of vectors**

You need to know that ***these formulas are not listed in the Physics data booklet.*** This means that in the examination you must be able either to write them down from memory or be able to derive them at the time. Most candidates find it easier to memorise them.

When a vector, *v*, is drawn at an angle θ to the horizontal, the horizontal component of that vector is '$v \cos \theta$', and the vertical component of the vector is '$v \sin \theta$', i.e.

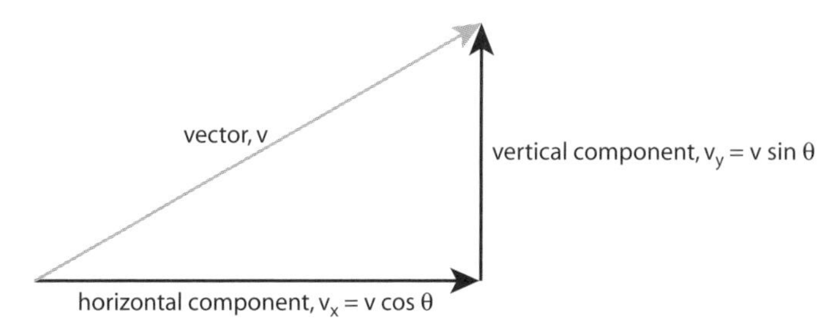

- **Inability to state or derive the formula for the component of weight down a slope**

You need to know that ***the formula is not listed in the Physics data booklet.*** This means that in the examination you must be able either to write it down from memory or be able to derive it at the time. Most candidates find it easier to memorise the formula.

When an object of mass, *m*, is on a slope at an angle of θ to the horizontal, the component of weight down the slope is $mg \sin \theta$, i.e.

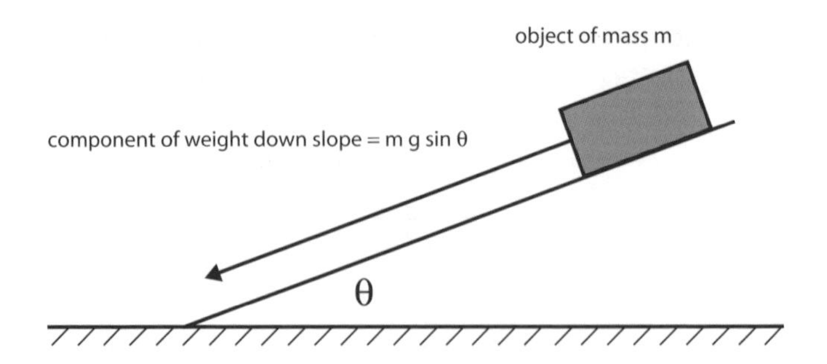

{'g' = the gravitational field strength of the planet (usually earth, where $g = 9{\cdot}8\ \text{N kg}^{-1}$)}

EXAM EXAMPLE 4

A crate of mass 40·0 kg is pulled up a slope using a rope.

The slope is at an angle of 30° to the horizontal.

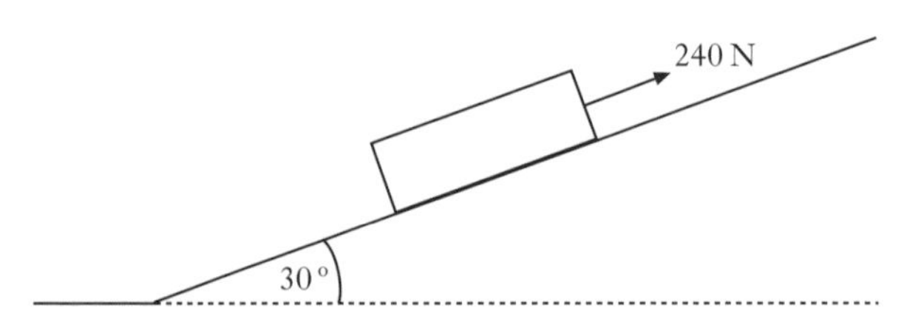

A force of 240 N is applied to the crate parallel to the slope.

The crate moves at a constant speed of $3{\cdot}0\ \text{m s}^{-1}$.

(*a*) (i) Calculate the component of the weight of the crate acting parallel to the slope.

component of weight = $mg \sin \theta$

$= 40{\cdot}0 \times 9{\cdot}8 \times \sin 30$

$= 196$ N

Equations of motion

- **Poor understanding of the meaning of the term 'acceleration'**

'Acceleration' means *how much the velocity changes each second*. By itself, the value of acceleration tells you nothing about the speed or the distance travelled.

EXAM EXAMPLE 5

An object has a constant acceleration of $3\,m\,s^{-2}$. This means that the

A distance travelled by the object increases by 3 metres every second

B displacement of the object increases by 3 metres every second

C speed of the object is $3\,m\,s^{-1}$ every second

D velocity of the object is $3\,m\,s^{-1}$ every second

E velocity of the object increases by $3\,m\,s^{-1}$ every second.

The answer is E, simply because this is the definition of what 'acceleration' means.

- **Incorrect substitution of initial velocity, *u*, and final velocity, *v***

You must remember that '*u*' stands for initial velocity and that '*v*' stands for final velocity. One method of remembering this is that 'initial' comes before 'final' and '*u*' comes before '*v*' in the alphabet.

The values must be substituted for the appropriate letters irrespective of which one is larger or whether they are positive or negative. For example:

EXAM EXAMPLE 6

To test the braking system of cars, a test track is set up as shown.

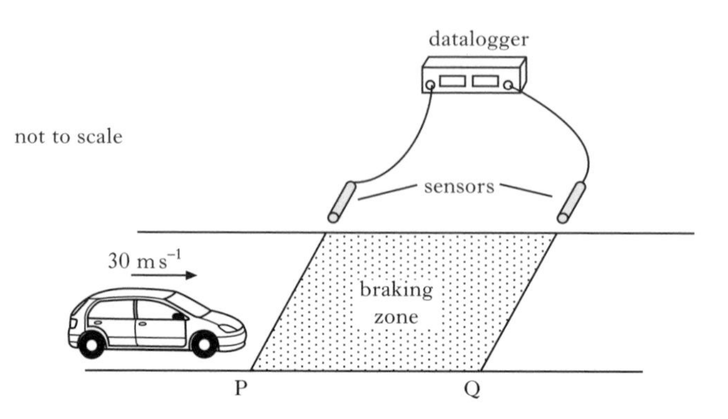

The sensors are connected to a datalogger which records the speed of a car at both P and Q.

A car is driven at a constant speed of 30 m s^{-1} until it reaches the start of the braking zone at P. The brakes are then applied.

(*a*) In one test, the datalogger records the speed at P as 30 m s^{-1} and the speed at Q as 12 m s^{-1}. The car slows down at a constant rate of 9·0 m s^{-2} between P and Q.

Calculate the length of the braking zone.

Answer: $u = 30$, $v = 12$, $a = -9{\cdot}0$, $s = ?$

$$v^2 = u^2 + 2as$$

$$12^2 = 30^2 + \{2 \times (-9{\cdot}0) \times s\}$$

$$\Rightarrow s = \frac{(144 - 900)}{-18} = \frac{-756}{-18} = 42 \text{ m}$$

- **Failing to substitute a *negative* value of acceleration when an object is slowing down**

Acceleration is a vector quantity and must be given a negative value if either a positive velocity is decreasing (as in the previous example) or when the unbalanced force is in the opposite direction to a positive initial velocity.

- **Inconsistent positive and negative signs for initial velocity, *u*, final velocity, *v*, and acceleration, *a***

In a similar way to the previous two issues, you need to remember that '*u*', '*v*' and '*a*' are all vector quantities. This means that their directions must be taken into consideration when substituting their values into an equation of motion.

You need to decide which direction you are going to consider as positive (up or down, right or left) and then be consistent in substituting any value as negative when it acts the opposite way.

EXAM EXAMPLE 7

3. A helicopter is **descending** vertically at a constant speed of $3{\cdot}0\,\text{m}\,\text{s}^{-1}$. A sandbag is released from the helicopter. The sandbag hits the ground 5·0 s later.

   What was the height of the helicopter above the ground at the time the sandbag was released?

   A 15·0 m

   B 49·0 m

   C 107·5 m

   D 122·5 m

   E 137·5 m

Taking *down* as positive:

$u = 3{\cdot}0,\ t = 5{\cdot}0,\ a = 9{\cdot}8,\ s = ?$

$s = ut + \frac{1}{2}at^2$

$s = (3{\cdot}0 \times 5{\cdot}0) + (\frac{1}{2} \times 9{\cdot}8 \times 5^2)$

$s = 15 + 122{\cdot}5$

$s = 137{\cdot}5$ m

OR

Taking *up* as positive:

$u = -3{\cdot}0$, $t = 5{\cdot}0$, $a = -9{\cdot}8$, $s = ?$

$s = ut + \frac{1}{2}at^2$

$s = (-3{\cdot}0 \times 5{\cdot}0) + (\frac{1}{2} \times -9{\cdot}8 \times 5^2)$

$s = -15 + (-122{\cdot}5)$

$s = -137{\cdot}5$ m

The negative sign means the displacement is 137·5 m *downwards*.

The answer is E.

- **Inability to draw an acceleration-time graph from a velocity-time graph**

The values of the acceleration should be calculated for each section of the velocity-time graph using $a = \frac{(v - u)}{t}$. These are then drawn as *horizontal* lines on an acceleration-time graph because acceleration is constant during each period of time.

EXAM EXAMPLE 8

(*b*) As the crate is moving up the slope, the rope snaps.

The graph shows how the velocity of the crate changes from the moment the rope snaps.

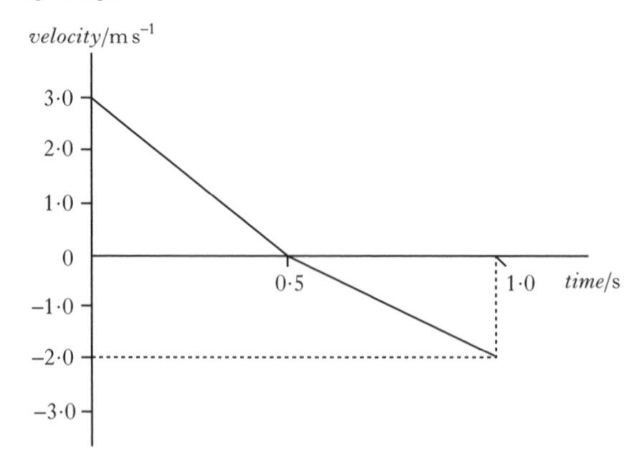

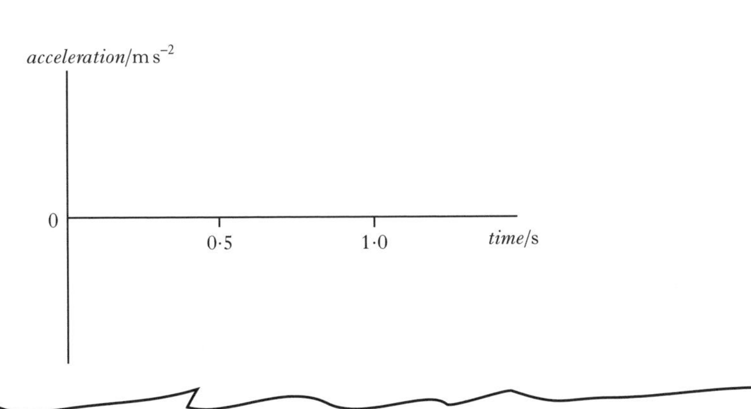
Copy the axes shown below and sketch the graph to show the acceleration of the crate between 0 and 1·0 s.

Appropriate numerical values are also required on the acceleration axis.

Answer:

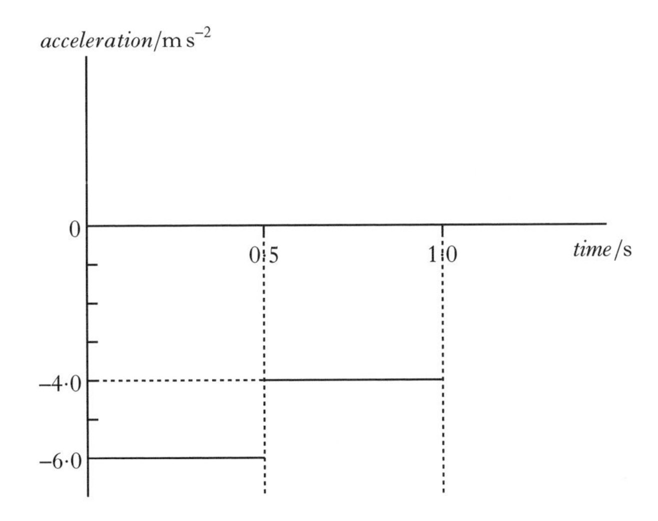

From $t = 0$ to $t = 0·5$ s,

$$a = \frac{(v - u)}{t}$$

$$a = \frac{(0 - 3)}{0·5}$$

$$a = -6·0 \text{ m s}^{-2}$$

Note: many candidates make the mistake of thinking the acceleration changes from zero to this value during the 0·5 seconds, but it remains constant at this value for the whole time.

From $t = 0{\cdot}5\ s$ to $t = 1{\cdot}0\ s$,

$$a = \frac{(v - u)}{t}$$

$$a = \frac{(-2{\cdot}0 - 0)}{0{\cdot}5}$$

$$a = -4{\cdot}0\ \text{m s}^{-2}$$

• Poor descriptions of an experimental method for measuring acceleration

One method to find acceleration is to find an initial velocity, u, a final velocity, v, and the time, t, for the change in velocity. The acceleration is then calculated from these values using $a = \frac{(v - u)}{t}$.

Possible experimental arrangements:

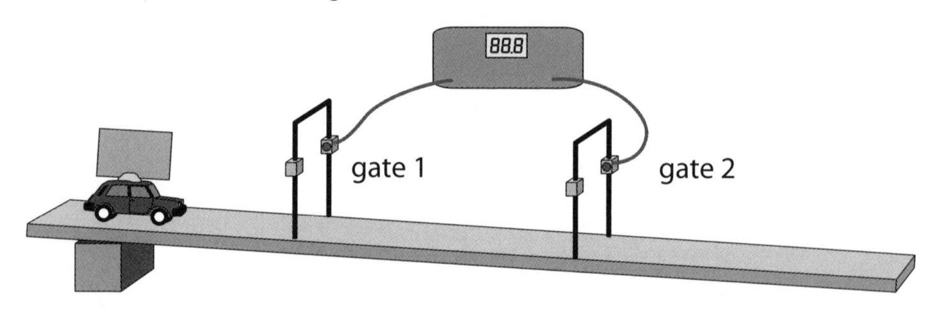

A single card with *two* light gates:

Gate 1 is used to find the initial velocity, u,
(u = length of card/time beam 1 is broken)

Gate 2 is used to find the final velocity, v,
(v = length of card/time beam 2 is broken)

The time, t, to travel from gate 1 to gate 2 must also be measured.

It is also possible to measure the acceleration of an object by using a single light gate and an object with a *double* card attached to it.

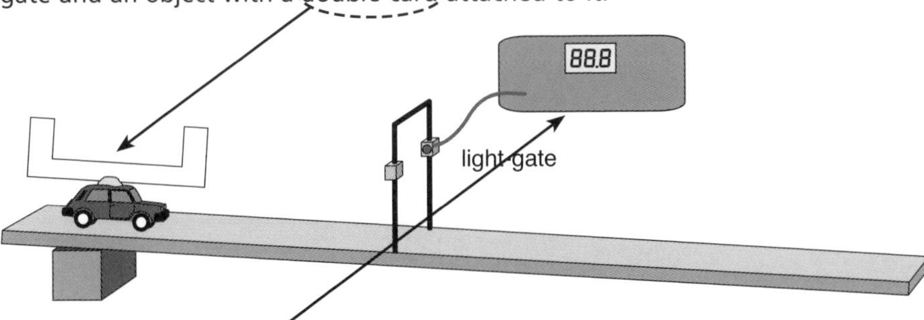

The attached electronic timer uses the different sections of the card to find the initial velocity, the final velocity and the time for the velocity change. It can then give a direct readout of the object's acceleration.

It is also possible to find acceleration by releasing an object from rest and measuring its final velocity after a measured displacement.

EXAM EXAMPLE 9

(*a*) A student uses the apparatus shown to measure the average acceleration of a trolley travelling down a track.

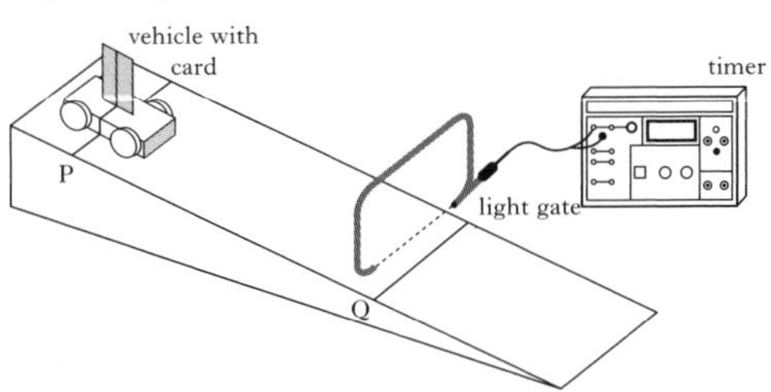

The line on the trolley is aligned with line P on the track.

The trolley is released from rest and allowed to run down the track.

The timer measures the time for the card to pass through the light gate.

This procedure is repeated a number of times and the results shown below.

0·015 s 0·013 s 0·014 s 0·019 s 0·017 s 0·018 s

(i) Calculate:

(A) the mean time for the card to pass through the light gate;

(B) the approximate absolute random uncertainty in this value.

(ii) The length of the card is 0·020 m and the distance PQ is 0·60 m.

Calculate the acceleration of the trolley (an uncertainty in this value is not required).

Answer:

(i) (A) 0·016 s
(B) 0·001 s

(ii) initial velocity, $u = 0$
displacement, $s = 0{\cdot}60$ m

$$\text{final velocity, } v = \frac{\text{length of card}}{\text{time}}$$

$$= \frac{0{\cdot}020}{0{\cdot}016}$$

$$= 1{\cdot}25\ \text{m s}^{-1}$$

acceleration, $a = ?$

$$v^2 = u^2 + 2as$$

$$1{\cdot}25^2 = 0^2 + (2 \times a \times 0{\cdot}60)$$

$$1{\cdot}5625 = 0 + 1{\cdot}2a$$

$$a = \frac{1{\cdot}5625}{1{\cdot}2}$$

$$a = 1{\cdot}30\ \text{m s}^{-2}$$

Newton's Second Law, Energy and Power

- **Poor understanding of the relationship between acceleration and velocity, especially in descriptive answers**

There is an overlap here with the earlier issue (poor understanding of the meaning of the term 'acceleration').

The definition of acceleration is 'the change in velocity each second'. This means that two different objects can have completely different speeds (or velocities) but have the same value of acceleration because *their velocities are changing at the same rate*.

For example, object A which speeds up from 2·5 m s^{-1} to 7·5 m s^{-1} in one second has the same acceleration as object B which speeds up from 312·5 m s^{-1} to 317·5 m s^{-1} in one second.

The acceleration of both objects is 5·0 metres per second per second (5·0 m s^{-2}).

(5·0 metres per second = the change in speed; per second = the time for the change)

EXAM EXAMPLE 10

(*b*) As the crate is moving up the slope, the rope snaps.

The graph shows how the velocity of the crate changes from the moment the rope snaps.

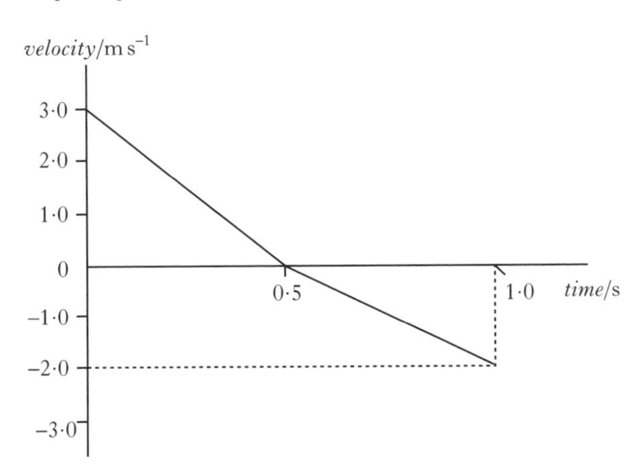

(i) Describe the motion of the crate during the first 0·5 s after the rope snaps.

Discussion:

For the first half second the velocity decreases from 3·0 m s^{-1} to zero at a constant rate.
This means the object has a constant negative acceleration.
Because the decrease in velocity is 3·0 m s^{-1} in half a second, the rate of change of velocity is −6·0 m s^{-1} per second, i.e.
The acceleration is −6·0 m s^{-2}.

Answer:

$$a = \frac{(v - u)}{t}$$

$$= \frac{(0 - 3{\cdot}0)}{0{\cdot}5} = -6{\cdot}0\ \text{m s}^{-2}$$

- **Poor ability to describe and explain how a number of forces combine to affect the acceleration or velocity of an object**

To improve your ability to answer such questions, first of all you need to ensure you understand Newton's second law, $F = ma$.

In the formula, '*F*' represents the *unbalanced force* acting on the object of mass m. This means that you need to combine all the forces acting on the object into one force. It is essential to take the directions of the forces into account when combining them, i.e. you need to allow for the fact that forces are vectors. Even when all of the forces act along a straight line it is a good idea to make a sketch of all of the forces acting in order to find the resultant force.

The acceleration of the object is always directly proportional to the unbalanced force and is in the same direction. The acceleration then tells you how rapidly the velocity is changing.

EXAM EXAMPLE 11

A fairground ride consists of rafts which slide down a slope into water.

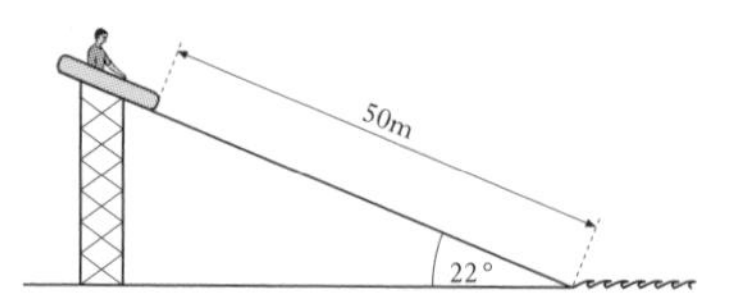

The slope is at an angle of 22° to the horizontal. Each raft has a mass of 8·0 kg. The length of the slope is 50 m.

A child of mass 52 kg sits in a raft at the top of the slope. The raft is released from rest. The child and raft slide together down the slope into the water. The force of friction between the raft and slope remains constant at 180 N.

(*a*) Calculate the component of weight, in newtons, of the child and raft down the slope.

(*b*) Show by calculation that the acceleration of the child and raft down the slope is $0{\cdot}67\ \text{m s}^{-2}$.

(*c*) Calculate the speed of the child and raft at the bottom of the slope.

(*d*) A second child of smaller mass is released from rest in an identical raft at the same starting point. The force of friction is the same as before.

How does the speed of this child and raft at the bottom of the slope compare with the answer to part (*c*)?

Justify your answer.

Answers:

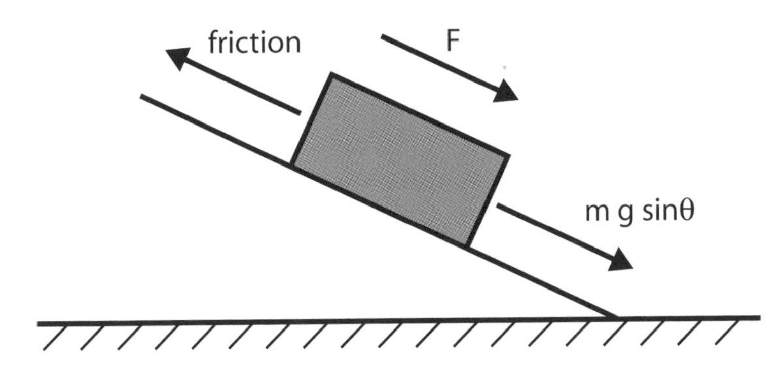

(a) $mg \sin \theta = (52 + 8) \times 9{\cdot}8 \times \sin 22$
$= 220$ N

(b) $F =$ component of weight − friction
$= 220 - 180 = 40$ N

$a = \frac{F}{m} = \frac{40}{60} = 0{\cdot}67\ \text{m s}^{-2}$

(c) $u = 0,\ a = 0{\cdot}67,\ s = 50,\ v = ?$

$v^2 = u^2 + 2as$
$= 0 + (2 \times 0{\cdot}67 \times 50) = 67$
$v = 8{\cdot}2\ \text{m s}^{-1}$

(d) smaller mass means the component of weight is smaller.
So the unbalanced force is smaller.
So the acceleration is smaller.
So the final velocity is smaller.

EXAM EXAMPLE 12

A car of mass 1200 kg pulls a horsebox of mass 700 kg along a straight, horizontal road. They have an acceleration of $2{\cdot}0\ \text{m s}^{-2}$.

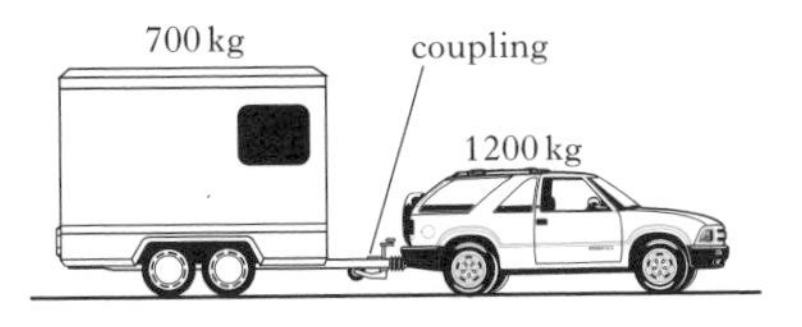

Assuming that the frictional forces are negligible, the tension in the coupling between the car and the horsebox is

A 500 N

B 700 N

C 1400 N

D 2400 N

E 3800 N.

Answer:

The tension in the coupling is the unbalanced force which is acting on the horsebox and causing it to accelerate.

Apply Newton's second law *to the horsebox*. The unbalanced force causing it to accelerate at $2{\cdot}0\ \mathrm{m\,s^{-2}}$ is

$F = ma$
$= 700 \times 2{\cdot}0$
$= 1400$ N (to the right)

The answer is C. (The mass of the car is irrelevant for this calculation.)

EXAM EXAMPLE 13

Two boxes on a frictionless horizontal surface are joined together by a string. A constant horizontal force of 12 N is applied as shown.

2·0 kg — 4·0 kg → 12 N

The tension in the string joining the two boxes is

A 2·0 N
B 4·0 N
C 6·0 N
D 8·0 N
E 12 N.

Answer:

To answer this type of question you need to apply Newton's second law *twice*. Firstly, apply it to the *whole system* to find the acceleration. Then apply it again to a *part of the system* to find the unbalanced force on that part.

For the whole system:

$F = ma$
$12 = 6{\cdot}0 \times a$
$a = 2{\cdot}0\ \mathrm{m\,s^{-2}}$ (to the right)

The tension in the string provides the force to accelerate the 2·0 kg mass.

For the 2·0 kg mass:

$F = ma$
$F = 2{\cdot}0 \times 2{\cdot}0$
$F = 4{\cdot}0$ N (to the right)

The answer is B.

EXAM EXAMPLE 14

4. A skydiver of total mass 85 kg is falling vertically.

At one point during the fall, the air resistance on the skydiver is 135 N.

The acceleration of the skydiver at this point is

A $0{\cdot}6\,\mathrm{m\,s^{-2}}$

B $1{\cdot}6\,\mathrm{m\,s^{-2}}$

C $6{\cdot}2\,\mathrm{m\,s^{-2}}$

D $8{\cdot}2\,\mathrm{m\,s^{-2}}$

E $13{\cdot}8\,\mathrm{m\,s^{-2}}$.

Answer:

weight of skydiver = mg = 85 × 9·8 = 833 N (downwards)

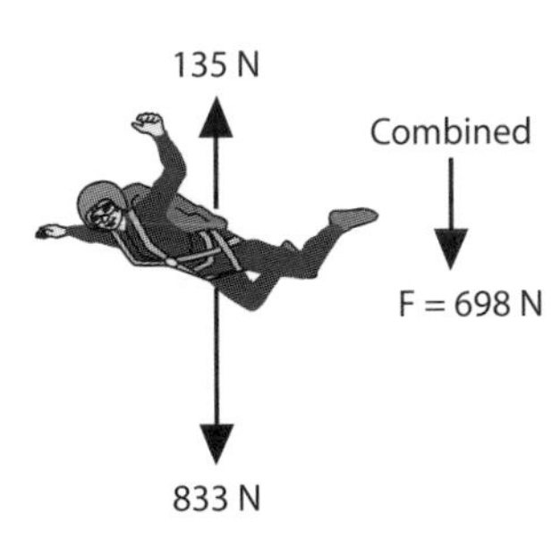

The unbalanced force on the skydiver is:

weight − air resistance = 833 − 135 = 698 N (down)

$a = \frac{F}{m} = \frac{698}{85} = 8{\cdot}21176\ \mathrm{ms^{-2}}$ (down)

The answer is D.

Momentum and Impulse

- **Lack of understanding of the vector nature of impulse and momentum and the link between these quantities**

Impulse = force × time.
Force requires a direction and so is a vector quantity – this makes impulse a vector quantity as well.

Momentum = mass × velocity.
Velocity requires a direction and so is a vector quantity – this makes momentum a vector quantity as well.

The link between momentum and impulse is:

impulse = change in momentum (= final momentum − initial momentum)
i.e. $Ft = mv - mu$

EXAM EXAMPLE 15

A golfer hits a ball of mass $5{\cdot}0 \times 10^{-2}$ kg with a golf club. The ball leaves the tee with a velocity of $80\,\text{m s}^{-1}$. The club is in contact with the ball for a time of 0·10 s.

The average force exerted by the club on the ball is

A $6{\cdot}25 \times 10^{-4}$ N

B 0·025 N

C 0·4 N

D 4 N

E 40 N.

Answer:

impulse = change in momentum
$Ft = m(v - u)$

So,
$F \times 0{\cdot}10 = 5{\cdot}0 \times 10^{-2} \times (80 - 0)$
$F \times 0{\cdot}10 = 4{\cdot}0$

$$F = \frac{4{\cdot}0}{0{\cdot}10}$$

$F = 40$ N

The answer is E.

EXAM EXAMPLE 16

23. Beads of liquid moving at high speed are used to move threads in modern weaving machines.

(*a*) In one design of machine, beads of water are accelerated by jets of air as shown in the diagram.

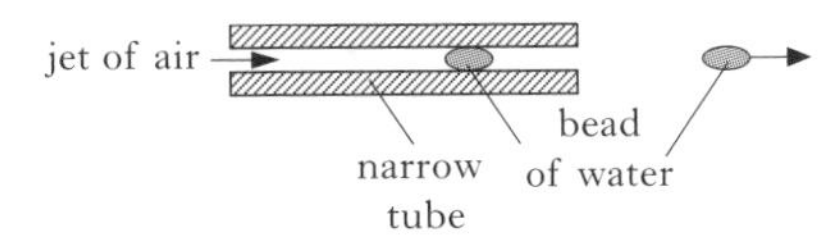

Each bead has a mass of $2{\cdot}5 \times 10^{-5}$ kg.

When designing the machine, it was estimated that each bead of water would start from rest and experience a constant unbalanced force of 0·5 N for a time of 3·0 ms.

(i) Calculate:

(A) the impulse on a bead of water;

(B) the speed of the bead as it emerges from the tube.

Answer:

(i)(A)

$$\text{impulse} = \text{force} \times \text{time}$$
$$= 0{\cdot}5 \times 3{\cdot}0 \times 10^{-3}$$
$$= 1{\cdot}5 \times 10^{-3}\ \text{Ns}$$

(i)(B)

$$\text{change in momentum} = \text{impulse}$$
$$m(v - u) = Ft$$

So,

$$m(v - u) = 1{\cdot}5 \times 10^{-3}$$

$$2{\cdot}5 \times 10^{-5}(v - 0) = 1{\cdot}5 \times 10^{-3}$$
$$v = \frac{1{\cdot}5 \times 10^{-3}}{2{\cdot}5 \times 10^{-5}}$$
$$= 60\ \text{ms}^{-1}$$

The change in momentum (= impulse) is equal to $mv - mu$ and you need to take great care both to substitute correctly for 'v' and 'u' and also to allow for the possibility of one of them being negative with respect to the other.

EXAM EXAMPLE 17

22. The apparatus shown below is used to test concrete pipes.

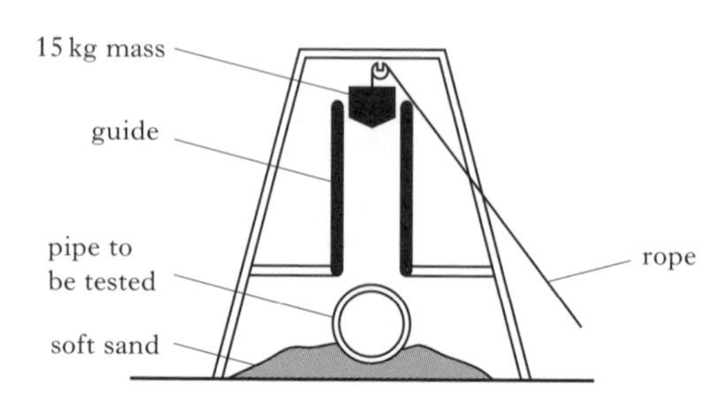

When the rope is released, the 15 kg mass is dropped and falls freely through a distance of 2·0 m on to the pipe.

(*a*) In one test, the mass is dropped on to an uncovered pipe.

(i) Calculate the speed of the mass just before it hits the pipe.

(ii) When the 15 kg mass hits the pipe the mass is brought to rest in a time of 0·02 s. Calculate the size and direction of the average unbalanced force on the **pipe**.

Answer:

(a) (i) Taking downwards as the positive direction:

$u = 0$, $a = 9{\cdot}8$, $s = 2{\cdot}0$, $v = ?$
$v^2 = u^2 + 2as$
$v^2 = 0^2 + (2 \times 9{\cdot}8 \times 2{\cdot}0) = 39{\cdot}2$
$v = 6{\cdot}26\ \mathrm{m\,s^{-1}} \Rightarrow$ The mass is travelling downwards at $6{\cdot}26\ \mathrm{m\,s^{-1}}$ just before it hits the pipe.

(a) (ii) Taking downwards as the positive direction:

When the mass collides with the pipe its initial velocity, u, is $6{\cdot}26\ \mathrm{m\,s^{-1}}$. After 0·02 *s* it comes to a stop and so its final velocity, v, is 0.
Impulse *on mass* = change in momentum

$$Ft = mv - mu$$
$$F \times 0{\cdot}02 = (15 \times 0) - (15 \times 6{\cdot}26)$$
$$F \times 0{\cdot}02 = -93{\cdot}9$$
$$F = \frac{-93{\cdot}9}{0{\cdot}02}$$

$F = -4695 = -4700$ N (rounding the answer to two significant figures)

The negative value for this answer shows that the *force on the mass is upwards.*

A candidate who substitutes the values for 'u' and 'v' the wrong way round gets an answer of $+4700$ N. However, this is regarded as wrong Physics at the substitution stage and most of the marks are lost.

Finally, to answer the question, the *pipe* experiences an equal and opposite force due to this collision and *so the unbalanced force on the pipe is 4700 N downwards.*

There are often questions using force-time graphs.

Impulse is equal to the area under a force-time graph. This means that you can calculate this area to give the change in momentum of an object when a force is applied to it.

EXAM EXAMPLE 18

The graph shows the force which acts on an object over a time interval of 8 seconds.

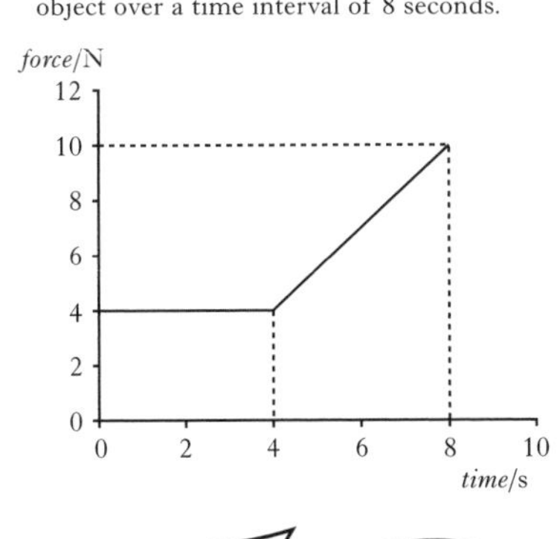

The momentum gained by the object during this 8 seconds is

A 12 kg m s^{-1}

B 32 kg m s^{-1}

C 44 kg m s^{-1}

D 52 kg m s^{-1}

E 72 kg m s^{-1}.

Answer:

gain in momentum = impulse
= area under force-time graph
= area A + area B

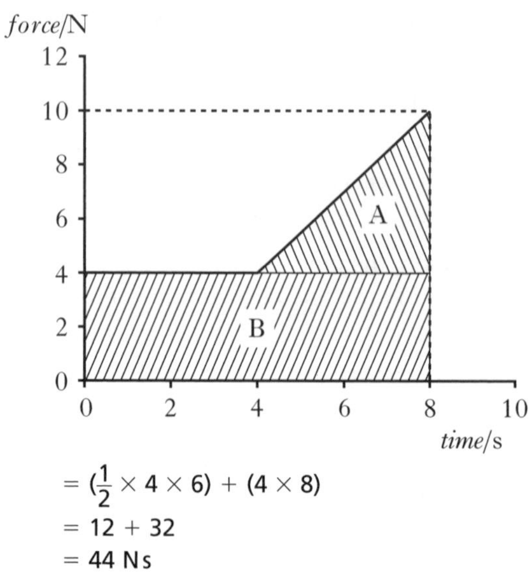

$= (\frac{1}{2} \times 4 \times 6) + (4 \times 8)$
$= 12 + 32$
$= 44 \text{ N s}$

OR 44 kg m s^{-1}

The answer is C.

- **Incorrect use of positive and negative signs for velocities in calculations**

The law of conservation of linear momentum is used to work out what happens in collisions. You must be very careful to decide which direction you wish to be positive (e.g. left to right). Anything moving the opposite way must then be given a *negative* velocity (and momentum).

EXAM EXAMPLE 19

Two trolleys travel towards each other in a straight line as shown.

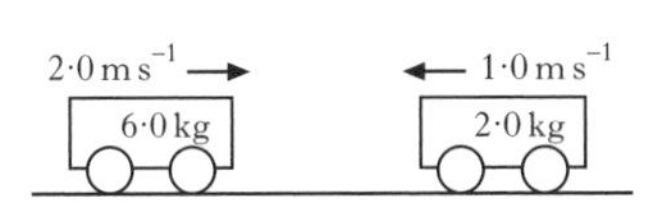

The trolleys collide. After the collision the trolleys move as shown below.

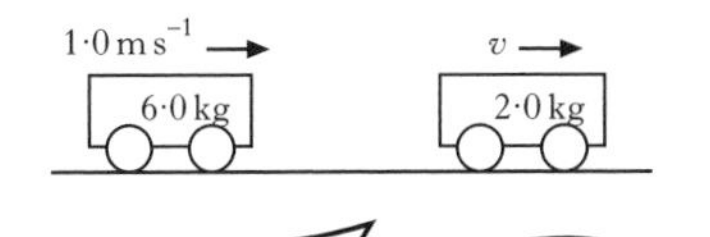

What is the speed v of the 2·0 kg trolley after the collision?

A $1{\cdot}25\ \text{m s}^{-1}$

B $1{\cdot}75\ \text{m s}^{-1}$

C $2{\cdot}0\ \text{m s}^{-1}$

D $4{\cdot}0\ \text{m s}^{-1}$

E $5{\cdot}0\ \text{m s}^{-1}$

Answer (taking left to right as the positive direction):

Total momentum before the collision = total momentum after the collision

$$m_1u_1 + m_2u_2 = m_1v_1 + m_2v_2$$

$$(6{\cdot}0 \times 2{\cdot}0) + (2{\cdot}0 \times -1{\cdot}0) = (6{\cdot}0 \times 1{\cdot}0) + (2{\cdot}0 \times v)$$

$$12 - 2 = 6 + 2v$$

$$2v = 4$$

$$v = 2\ \text{ms}^{-1}.$$

The answer is C.

- **Poor understanding of the meaning of an 'elastic' collision**

Many candidates make the mistake of thinking that an elastic collision means that two objects do not stick together. However, the correct definition of an elastic collision is *'one in which the total kinetic energy remains constant'*.

The following question involves most of the issues mentioned in this section about impulse and momentum.

EXAM EXAMPLE 20

Two ice skaters are initially skating together, each with a velocity of $2{\cdot}2\,\text{m s}^{-1}$ to the right as shown.

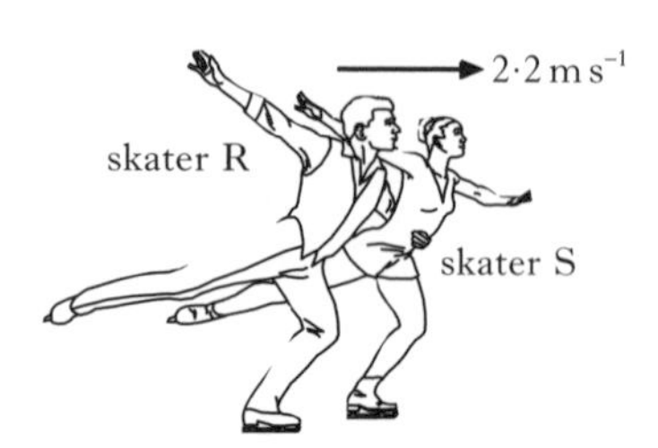

The mass of skater R is 54 kg. The mass of skater S is 38 kg.

Skater R now pushes skater S with an average force of 130 N for a short time. This force is in the same direction as their original velocity.

As a result, the velocity of skater S increases to $4{\cdot}6\,\text{m s}^{-1}$ to the right.

(*a*) Calculate the magnitude of the change in momentum of skater S.

(*b*) How long does skater R exert the force on skater S?

(*c*) Calculate the velocity of skater R immediately after pushing skater S.

(*d*) Is this interaction between the skaters elastic?

You must justify your answer by calculation.

Answer:

(a) change in momentum = final momentum − initial momentum = $mv - mu$

$$= (38 \times 4{\cdot}6) - (38 \times 2{\cdot}2)$$
$$= 174{\cdot}8 - 83{\cdot}6$$
$$= 91{\cdot}2\ \text{kg m s}^{-1}$$

(b) impulse = Ft = change in momentum

$130 \times t = 91{\cdot}2$

$$t = \frac{91{\cdot}2}{130} = 0{\cdot}70 \text{ s}$$

(c)

$$m_1u_1 + m_2u_2 = m_1v_1 + m_2v_2$$

$$(54 \times 2{\cdot}2) + (38 \times 2{\cdot}2) = (54 \times v) + (38 \times 4{\cdot}6)$$

$$54\,v = 118{\cdot}8 + 83{\cdot}6 - 174{\cdot}8\ (=27{\cdot}6)$$

$54\,v = 27{\cdot}6$ which gives $v = \frac{27{\cdot}6}{54} = 0{\cdot}51 \text{ m s}^{-1}$.

(d) total kinetic energy before $= \frac{1}{2}mu^2 = \frac{1}{2}(54 + 38) \times 2{\cdot}2^2 = 223$ J

total kinetic energy after $= \frac{1}{2}m_1v_1^{\,2} + \frac{1}{2}m_2v_2^{\,2} = (\frac{1}{2}\,54 \times 0{\cdot}51^2) + (\frac{1}{2}\,38 \times 4{\cdot}6^2) = 409$ J

Total kinetic energy is not constant and so this is *not an elastic collision.*

Density and Pressure

- **Lack of ability to use the density formula to describe/explain how the density of a sample of gas is affected when environmental factors change**

Density is defined as mass per unit volume and can be calculated using:

$$\text{density} = \frac{\text{mass}}{\text{volume}}$$

Candidates regularly show that they can use this formula to perform calculations correctly, but they cannot use it to explain the behaviour of a gas when its volume changes.

In your answer, you should:

- quote the relationship
- state what happens to the mass (even when it remains constant)
- state what happens to the volume (even when it remains constant)
- work out whether the ratio of 'mass/volume' increases, decreases or stays the same

EXAM EXAMPLE 21

A cylinder of compressed oxygen gas is in a laboratory.

(*a*) The oxygen inside the cylinder is at a pressure of $2{\cdot}82 \times 10^6$ Pa and a temperature of 19·0 °C.

The cylinder is now moved to a storage room where the temperature is 5·0 °C.

(ii) What effect, if any, does this decrease in temperature have on the density of the oxygen in the cylinder?

Justify your answer.

Answer to (a)(ii):

The mass of gas in the cylinder is constant (as none can enter or escape).

The volume of the cylinder does not change (as it is rigid and the small change in temperature will have little effect).

$$\text{density} = \frac{\text{mass}}{\text{volume}}$$

the ratio of $\frac{m}{V}$ is constant and so the *density does not change.*

EXAM EXAMPLE 22

6. The density of the gas in a container is initially $5{\cdot}0\,\text{kg}\,\text{m}^{-3}$.

Which of the following increases the density of the gas?

I Raising the temperature of the gas without changing its mass or volume.

II Increasing the mass of the gas without changing its volume or temperature.

III Increasing the volume of the gas without changing its mass or temperature.

A II only

B III only

C I and II only

D II and III only

E I, II and III

density, $\rho = \dfrac{m}{V}$

I $\rightarrow$ ρ constant

II $\rightarrow$ ρ increases

III $\rightarrow$ ρ decreases

The answer is A.

- **Poor attempts at explaining buoyancy force in terms of the pressure difference between the top and bottom of an object**

The full explanation for buoyancy force (or upthrust) is as follows:

- pressure increases with depth, h, according to the relationship $P = \rho gh$
- the pressure on the bottom surface, P_b, is therefore greater than the pressure on the top surface, P_t
- Force, F, is directly proportional to the pressure, P, according to the relationship $F = PA$
- the upward force on the bottom surface, F_b, is greater than the downward force on the top surface, F_t, because $P_b > P_t$
- there is therefore an unbalanced force, F_U, acting upwards on the object – this unbalanced force is called 'upthrust'.

Diagramatically:

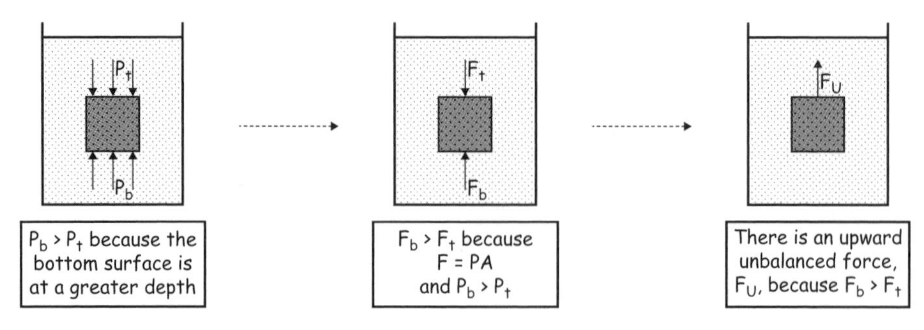

It is hoped that you can understand each step of this argument.
However, you may find it necessary to keep writing it out until it is fixed in your memory.

Gas Laws

- **Failure to change temperatures into Kelvins before substituting**

This is a common error which results in the loss of nearly all marks.

It is essential to realise that the gas laws are established on the basis of temperatures being in Kelvin units. **The Gas Laws do not work when temperatures are in Celsius degrees.**

To change from a temperature in Celsius to Kelvins, add 273. For example, the air in a room at 23°C is at a temperature of (23 + 273) = 296 K.

The value of '273' has to be memorised – it is not listed in the Physics data booklet.

EXAM EXAMPLE 23

A cylinder of compressed oxygen gas is in a laboratory.

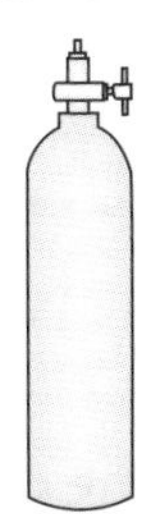

(*a*) The oxygen inside the cylinder is at a pressure of $2 \cdot 82 \times 10^6$ Pa and a temperature of 19·0 °C.

The cylinder is now moved to a storage room where the temperature is 5·0 °C.

(i) Calculate the pressure of the oxygen inside the cylinder when its temperature is 5·0 °C.

Answer:

$T_1 = 19°C = (19 + 273)$ K
$= 292$ K

$T_2 = 5°C = (5 + 273)$ K
$= 278$ K

$$\frac{P_1}{T_1} = \frac{P_2}{T_2}$$

$$\frac{2 \cdot 82 \times 10^6}{292} = \frac{P_2}{278}$$

$$P_2 = 278 \times \frac{2 \cdot 82 \times 10^6}{292}$$

$$P_2 = 2 \cdot 68479 \times 10^6$$
$$= 2 \cdot 68 \times 10^6 \text{ Pa}$$

(rounding to 3 sig. figs)

- **In an explanation of why a gas exerts a pressure, lack of reference to the collisions of the molecules with the walls of the container**

The molecules in any sample of gas are in constant, random motion. This means that the molecules of the gas are moving around in all directions with a range of different speeds. Any one molecule keeps moving at *constant speed in a straight line* until it has a collision with the walls or another molecule.

It is the gas *molecules colliding with the walls of the container* which causes a force (and pressure) on the walls.

Your answer must refer to these collisions *with the walls* to explain the pressure caused by the gas.

For an example, see the question given after the next issue.

- **In an explanation of why gas pressure changes, lack of reference to any change in the frequency of the collisions of the molecules with the walls of the container and/or the force of each collision**

There are two separate factors which affect the pressure.

The first factor is how hard each collision is. For example, a faster moving molecule collides harder with the wall and so exerts a greater force.

The second factor is how many molecules are hitting the walls each second – i.e. the frequency of the collisions. A greater number of collisions per second also increases the force on the walls.

Your answer must refer to which of these factors are relevant in a given question.

EXAM EXAMPLE 24

23. A student is training to become a diver.

(*a*) The student carries out an experiment to investigate the relationship between the pressure and volume of a fixed mass of gas using the apparatus shown.

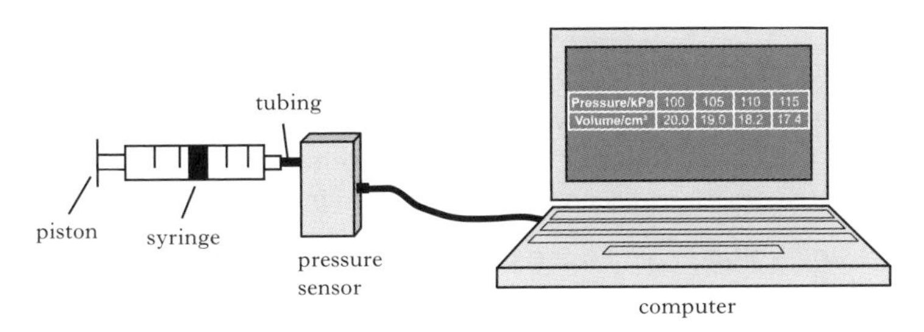

The pressure of the gas is recorded using a pressure sensor connected to a computer. The volume of the gas is also recorded. The student pushes the piston to alter the volume and a series of readings is taken.
The temperature of the gas is constant during the experiment.

The results are shown.

Pressure/kPa	100	105	110	115
Volume/cm^3	20·0	19·0	18·2	17·4

(ii) Use the kinetic model to explain the change in pressure as the volume of gas decreases.

Answer to (a)(ii):

As the volume is decreased, the same number of molecules have less room in which to move around. This means they strike the walls of the container more frequently. This greater number of collisions per second increases the average force on the walls. As a result, the pressure exerted by the gas increases.

Note: In this question it would be wrong Physics to state that the individual collisions are harder – this is because the temperature has not changed and so the molecules are still moving at the same speed.

- **Poor attempts at using the kinetic model to qualitatively explain the gas laws**

This is closely related to the last two issues. You need to be able to explain the following three relationships in terms of changes in the collisions of the gas molecules with the walls of the container.

Pressure and Temperature: When the temperature is increased, the kinetic energy of the molecules increases. This means that the molecules move faster. As a result, the collisions with the walls are harder and more frequent. Each of these two effects increases the force on the walls and therefore the pressure increases.

Pressure and Volume: When the volume of the container is decreased there is less space for the molecules to move around in. As a result, they collide with the walls more frequently (but not harder, as the temperature is constant). This increases the force on the walls and so the pressure increases.

Temperature and Volume: When the temperature of the gas is increased the molecules move around faster (greater E_k). As a result, each collision with the walls is harder. However, the force and pressure on the walls remains constant. There must therefore be fewer collisions per second. The only way of achieving this is for the molecules to move further apart – the volume increases.

- **Lack of understanding that a cylinder of gas is not completely 'emptied' when gas is released**

When the gas in a cylinder is allowed to expand to a greater volume at a lower pressure *some of the gas remains within the cylinder*, i.e.

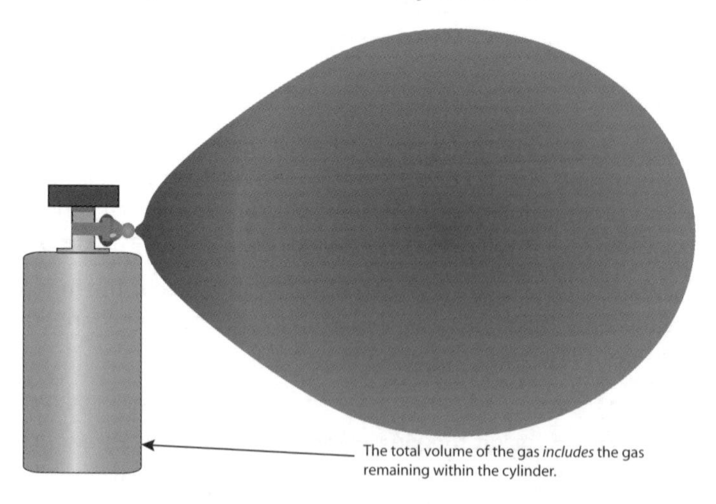

EXAM EXAMPLE 25

A rigid cylinder contains 8·0 ˘ 10^{-2} m^3 of helium gas at a pressure of 750 kPa. Gas is released from the cylinder to fill party balloons.

During the filling process, the temperature remains constant. When filled, each balloon holds 0·020 m^3 of helium gas at a pressure of 125 kPa.

(*a*) Calculate the total volume of the helium gas when it is at a pressure of 125 kPa.

(*b*) Determine the maximum number of balloons which can be fully inflated by releasing gas from the cylinder.

Answer:

(a)

$$P_1V_1 = P_2V_2$$
$$750 \times 8{\cdot}0 \times 10^{-2} = 125 \times V_2$$
$$V_2 = 750 \times \frac{8{\cdot}0 \times 10^{-2}}{125}$$
$$V_2 = 48 \times 10^{-2}\ m^3$$

(b) volume of gas which escapes to fill balloons is equal to the total volume at the new pressure minus the volume which remains inside the cylinder.

volume of gas available to fill balloons $= 0{\cdot}48 - 0{\cdot}08$
$= 0{\cdot}40$

$$\text{Number of balloons} = \frac{0{\cdot}40}{0{\cdot}020}$$
$$= 20$$

ADDRESSING AREAS OF WEAKNESS IN ELECTRICITY AND ELECTRONICS

Electric Fields and Resistors in Circuits

- **Inability to define the term 'potential difference'**

Content statements 2.1.4 and 2.1.5 say that you should be able to "State that the potential difference between two points is a measure of the work done in moving one coulomb of charge between the two points" and "State that if one joule of work is done moving one coulomb of charge between two points, the potential difference between the two points is one volt".

You need to learn these well enough to be able to write them out from memory. You also need to be able to adapt them to include appropriate values for a particular question.

EXAM EXAMPLE 26

The diagram below shows the basic features of a proton accelerator. It is enclosed in an evacuated container.

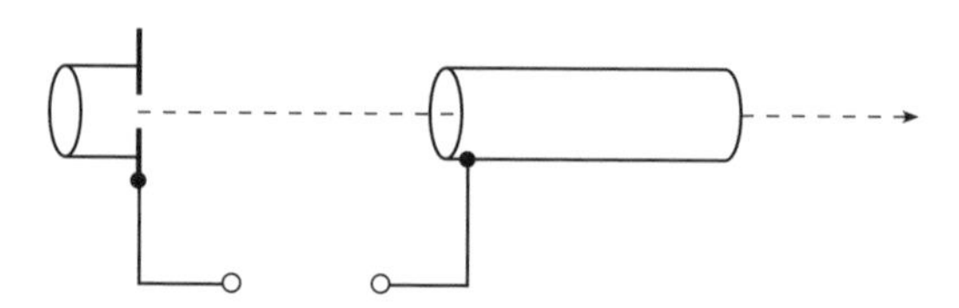

Protons released from the proton source start from rest at **P**.
A potential difference of 200 kV is maintained between **P** and **Q**.

(*a*) What is meant by the term *potential difference of 200 kV*?

Answer:

Each coulomb of charge gains 200 000 J of energy as it moves from P to Q.

- **Inability to define the term 'e.m.f.'**

Content statement 2.1.7 says that you should be able to "State that the e.m.f. of a source is the electrical potential energy supplied to each coulomb of charge which passes through the source".

You need to learn this well enough to be able to write it out from memory. You also need to be able to adapt it to include appropriate values for a particular question.

EXAM EXAMPLE 27

A battery has an e.m.f. of 6·0 V and internal resistance of 2·0 Ω.

(*a*) What is meant by an *e.m.f. of 6·0 V*?

Answer:

Each coulomb of charge gains 6·0 J of electrical potential energy as it passes through the battery/source.

- **Confusion between the two formulas for calculating energy, '*QV*' and '$\frac{1}{2}QV$' (see Capacitance)**

When a charge, *Q*, is moved through a *constant* potential difference of '*V*' volts, the work done, *W*, is calculated from $W = QV$.

EXAM EXAMPLE 28

A potential difference of 5000 V is applied between two metal plates. The plates are 0·10 m apart. A charge of +2·0 mC is released from rest at the positively charged plate as shown.

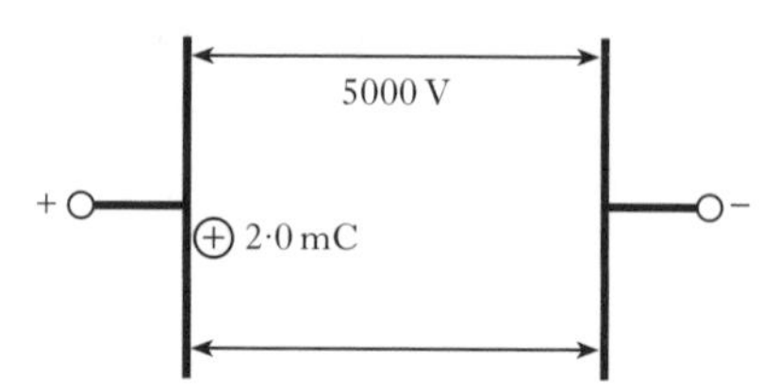

The kinetic energy of the charge just before it hits the negative plate is

A $4{\cdot}0 \times 10^{-7}$ J

B $2{\cdot}0 \times 10^{-4}$ J

C 5·0 J

D 10 J

E 500 J.

Answer:

Gain in kinetic energy = work done on charge $Q = QV$
$= 2{\cdot}0 \times 10^{-3} \times 5000$
$= 10$ J

The answer is D.

When charge Q is transferred from one plate of a capacitor to the other plate, the potential difference between the plates is *not constant*, but increases from zero to 'V' volts. The average potential difference during this charging process is therefore '$\frac{1}{2}V$' and the work done is $\frac{1}{2}QV$.

EXAM EXAMPLE 29

An uncharged 2200 μF capacitor is connected in a circuit as shown.

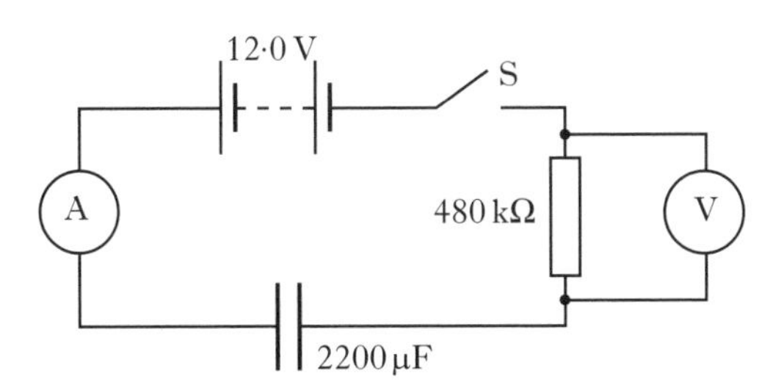

The battery has negligible internal resistance.

(c) Calculate the **maximum** energy the capacitor stores in this circuit

Answer:

The maximum p.d. across the capacitor in this circuit is 12 V.

$$\text{Energy stored} = \frac{1}{2}QV$$
$$= \frac{1}{2}CV\,V \quad \text{OR} \quad \frac{1}{2}CV^2$$
$$= \frac{1}{2} \times 2200 \times 10^{-6} \times 12^2$$
$$= 0{\cdot}1584 \text{ J}$$

The simple way to decide which formula to use is to remember that the formula with the '$\frac{1}{2}$' is used for questions where a capacitor is being charged.

- **Poor understanding of current and potential differences in circuits containing a number of resistors in series and parallel**

Series circuits and parallel circuits 'behave' in different ways.

In series circuits, the current is the same everywhere (and is determined by the supply voltage divided by the total resistance of the circuit). The supply voltage divides up across the series resistors – the voltage across each resistor is in proportion to its fraction of the total resistance.

In parallel circuits the current from the supply is also calculated from $\frac{V_{supply}}{R_{total}}$, but this current splits into the different parallel paths – the greater the resistance of a particular path, the smaller the current in it. The voltage is the same across all components which are in parallel with each other.

Ohm's law ($V = IR$) applies in all circuits. It is often useful to apply this relationship to the whole circuit (e.g. using total resistance and supply voltage to calculate current) and then to use it again applied to part of the circuit (e.g. using circuit current and a resistance value to calculate the voltage across that resistor).

EXAM EXAMPLE 30

A battery of e.m.f. 12 V and internal resistance 3·0 Ω is connected in a circuit as shown.

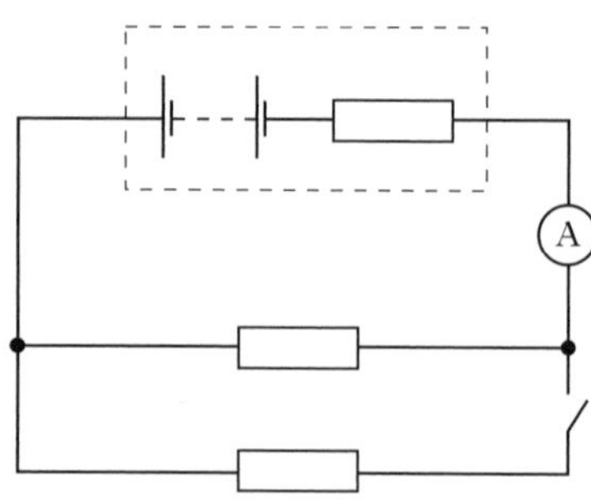

When switch **S** is closed the ammeter reading changes from

A 2·0 A to 1·0 A

B 2·0 A to 2·4 A

C 2·0 A to 10 A

D 4·0 A to 1·3 A

E 4·0 A to 6·0 A.

Answer:

One way of answering this question is as follows.

When *S* is open:

external resistance = 3·0 Ω
total circuit resistance = 3·0 + 3·0 = 6·0 Ω

$$\text{current} = \frac{V_{\text{total}}}{R_{\text{total}}} = \frac{12}{6{\cdot}0} = 2{\cdot}0\text{ A}$$

When *S* is closed:

the external resistance is the parallel combination of 3·0 Ω and 6·0 Ω

$$\frac{1}{R_t} = \frac{1}{R_1} + \frac{1}{R_2}$$

$$= \frac{1}{3} + \frac{1}{6} = \frac{3}{6}$$

$$\Rightarrow R_t = \frac{6}{3} = 2{\cdot}0\ \Omega$$

total circuit resistance = 2·0 + 3·0 = 5·0 Ω

$$\text{current} = \frac{V_{\text{total}}}{R_{\text{total}}} = \frac{12}{5{\cdot}0} = 2{\cdot}4\text{ A}$$

The answer is B.

Alternating Current and Voltage

- **Misreading/misuse of displays of the settings on an oscilloscope**

The y-gain setting on an oscilloscope controls the *height* of the displayed signal. This setting tells you the number of volts (or millivolts) for each division up and down the screen.

The amplitude of an alternating signal is the distance from the zero line to the top of the trace, i.e.

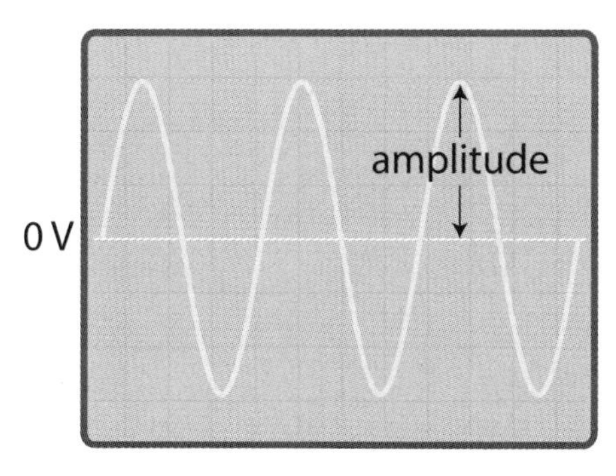

A y-gain setting of 5·0 V/div produces a smaller amplitude than one of 2·0 V/div. The y-gain setting is multiplied by the number of divisions of the amplitude in order to calculate the peak voltage of the signal. The r.m.s. voltage is then calculated by dividing the peak voltage by √2.

The timebase setting on an oscilloscope controls the *width* of a wave on the screen. This setting gives the number of seconds (or milliseconds) for each division across the screen. A setting of 5·0 ms/div produces more waves across the screen than a setting of 2·0 ms/div.

The timebase setting is multiplied by the number of divisions across the screen for one wave in order to calculate the period, T, of the signal. The frequency, f, of the signal can then be calculated from $f = \frac{1}{T}$.

EXAMPLE 31

An alternating voltage is applied to the input of an oscilloscope. The diagram below shows the trace displayed

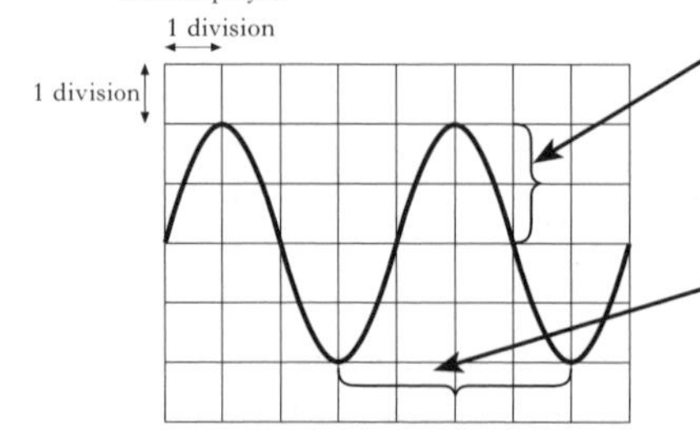

The Y-gain is set at 20 V/division. The timebase is set at 10ms/division.

Identify the row in the table below that shows the peak voltage and the frequency of the signal.

	Peak voltage/V	*Frequency*/Hz
A	14·2	10
B	28	25
C	40	10
D	40	25
E	80	25

Answer:

The amplitude of this trace is 2 divisions.

The peak voltage is therefore $2 \times 20\ \text{V} = 40\ \text{V}$

Horizontally, one wave occupied 4 divisions.

The period of a wave is therefore $4 \times 10\ \text{ms} = 40\ \text{ms}$.

The frequency,

$$f = \frac{1}{T} = \frac{1}{(40 \times 10^{-3})} = 25\ \text{Hz}$$

The answer is D.

EXAM EXAMPLE 32

A signal from a power supply is displayed on an oscilloscope.

The trace on the oscilloscope is shown.

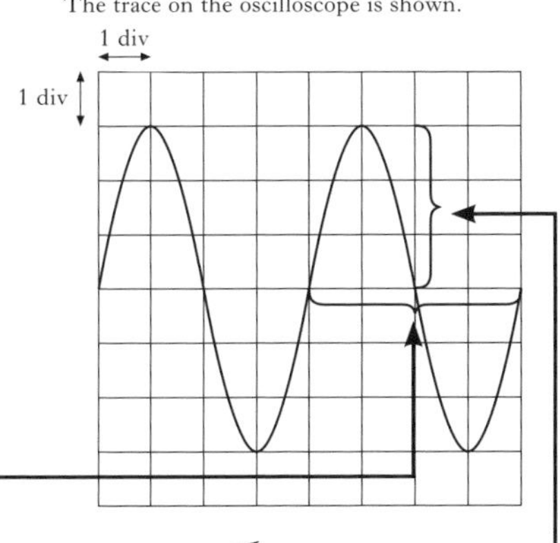

The time-base is set at 0·01 s/div and the Y-gain is set at 4·0 V/div.

Which row in the table shows the r.m.s. voltage and the frequency of the signal?

	r.m.s. voltage/V	*frequency*/Hz
A	8·5	25
B	12	25
C	24	25
D	8·5	50
E	12	50

Answer:

The amplitude of this trace is 3 divisions.

The peak voltage is therefore $3 \times 4{\cdot}0\ \text{V} = 12\ \text{V}$

The r.m.s. voltage is $\frac{V_p}{\sqrt{2}} = \frac{12}{\sqrt{2}} = 8{\cdot}5\ \text{V}$

Horizontally, one wave occupied 4 divisions.

The period of a wave is therefore $4 \times 0{\cdot}01\ \text{s} = 0{\cdot}04\ \text{s}$.

The frequency, $f = \frac{1}{T} = \frac{1}{0{\cdot}04} = 25\ \text{Hz}$

The answer is A.

- **Lack of knowledge of the relationship between current and frequency in a resistive circuit**

The current is the same at all frequencies in a resistive circuit. This is often shown by a current-frequency graph. The graph is a straight, horizontal line.

EXAM EXAMPLE 33

A resistor is connected in series with an a.c. supply and an ammeter as shown in the diagram below.

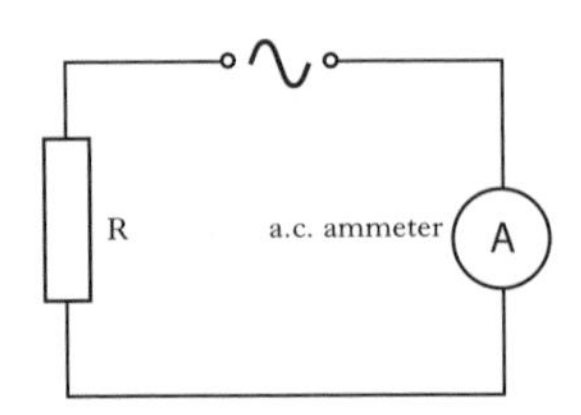

The frequency of the suply is varied while its peak voltage remains constant. As the frequency is steadily increased, the reading on the ammeter

A stays constant

B steadily decreases

C steadily increases

D decreases then increases

E increases then decreases.

The answer is A.

EXAM EXAMPLE 34

A resistor and an ammeter are connected to a signal generator which has an output of constant amplitude and variable frequency.

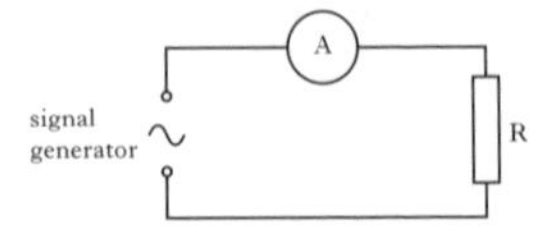

Which graph shows the relationship between the current I in the resistor and the output frequency f of the signal generator?

A

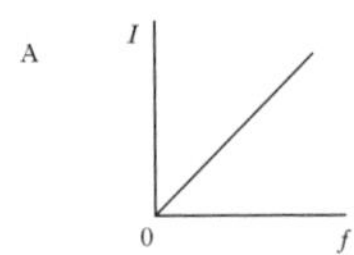

B

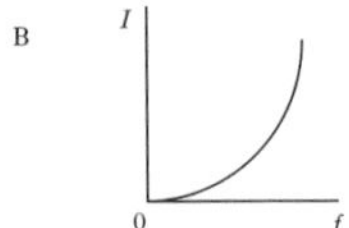

C

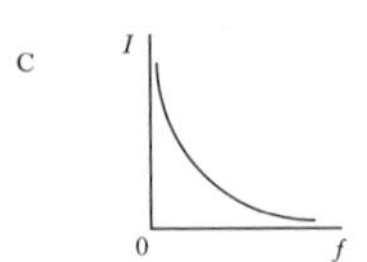

D

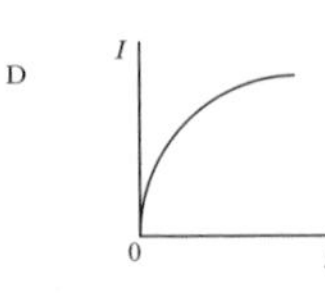

E

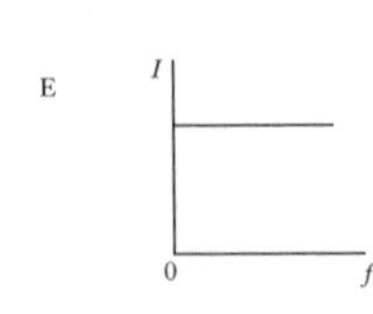

The answer is E.

Capacitance

- **Poor understanding of the processes of charging and discharging a capacitor**

When an uncharged capacitor is connected in series to a power supply and a resistor, electrons flow from the negative terminal of the supply on to one plate of the capacitor. At the same time, electrons are attracted off the other plate of the capacitor and flow to the positive terminal of the supply. Electrons continue to flow in the circuit until the potential difference across the plates of the capacitor is equal in size to the power supply voltage. The greater the value of the series resistor, the longer the time the charging process takes. During the charging process, the current, I, in the circuit decreases to zero and the p.d. across the plates of the capacitor *increases*.

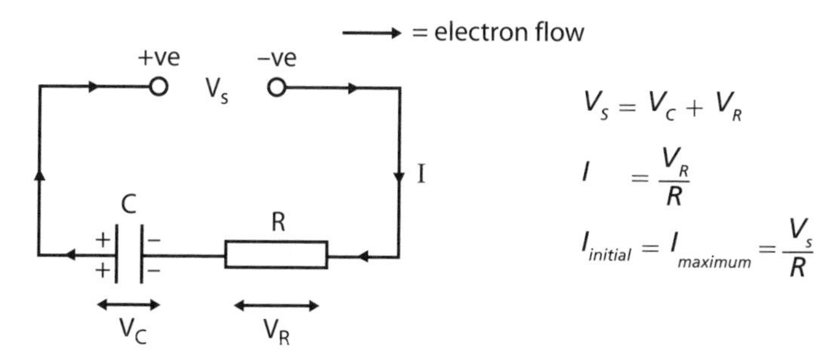

$$V_S = V_C + V_R$$

$$I = \frac{V_R}{R}$$

$$I_{initial} = I_{maximum} = \frac{V_s}{R}$$

When a charged capacitor is discharged through a series resistor, electrons flow round the circuit from the negative plate to the positive plate reducing the stored charge. The greater the value of the series resistor, the longer the time the discharging process takes. During the discharging process, the current in the circuit decreases to zero and the p.d. across the capacitor *decreases*.

EXAM EXAMPLE 35

(*a*) State what is meant by the term *capacitance*.

(*b*) An uncharged capacitor, C, is connected in a circuit as shown.

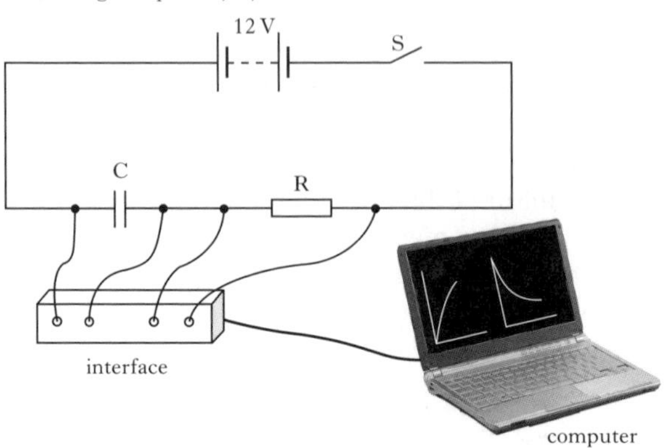

The 12 V battery has negligible internal resistance.

Switch S is closed and the capacitor begins to charge.

The interface measures the current in the circuit and the potential difference (p.d.) across the capacitor. These measurements are displayed as graphs on the computer.

Graph 1 shows the p.d. across the capacitor for the first 0·40 s of charging.

Graph 2 shows the current in the circuit for the first 0·40 s of charging.

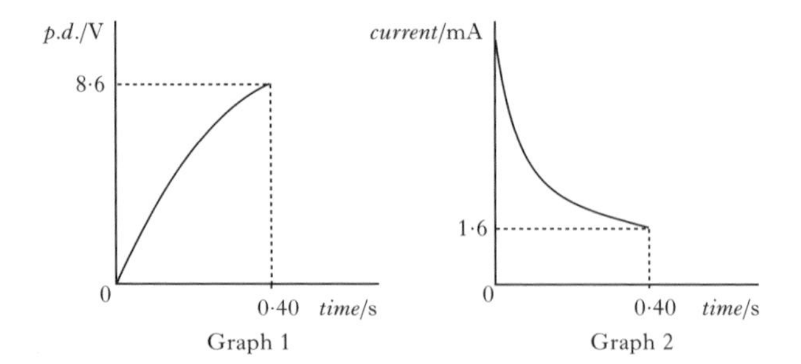

Graph 1 Graph 2

(i) Determine the p.d. **across resistor R** at 0·40 s.

(ii) Calculate the resistance of R.

(iii) The capacitor takes 2·2 seconds to charge fully.

At that time it stores 10·8 mJ of energy.

Calculate the capacitance of the capacitor.

Answer:

(a) Capacitance = the number of coulombs of charge stored for every volt of potential difference across the plates of a capacitor.

(b) (i) $V_R = V_S - V_C = 12 - 8{\cdot}6 = 3{\cdot}4$ V

(ii) $R = \frac{V_R}{I} = \frac{3{\cdot}4}{1{\cdot}6 \times 10^{-3}} = 2125\ \Omega.$

(iii) When fully charged, in this circuit, the p.d. across the capacitor $= V_S = 12\ V$

energy $= E_C = 10{\cdot}8 \times 10^{-3}$ J $= \frac{1}{2}CV^2$ so $\frac{1}{2}CV^2 = 10{\cdot}8 \times 10^{-3}$

$\Rightarrow C = 2 \times \frac{10{\cdot}8 \times 10^{-3}}{V^2}$

$\Rightarrow C = 2 \times \frac{10{\cdot}8 \times 10^{-3}}{144}$

$\Rightarrow C = 1{\cdot}5 \times 10^{-4}$ F

- **Confusion between the two formulas for calculating energy, 'QV' and '$\frac{1}{2}QV$' (see also Electric Fields and Resistors in Circuits)**

When charge Q is transferred from one plate of a capacitor to the other plate, the potential difference between the plates is not constant, but increases from zero to 'V' volts. The average potential difference during charging is therefore '$\frac{1}{2}V$' and the work done is $\frac{1}{2}QV$.

A simple way to decide which of the two formulas to use is to remember that the formula with the '$\frac{1}{2}$' is used for questions where a capacitor is being charged.

- **Poor ability to analyse current, potential differences and stored energy when a capacitor and resistor are connected in series with a d.c. supply**

The sum of the potential differences across the capacitor and resistor equals the supply voltage, i.e. $V_S = V_C + V_R$.

The current in the circuit, $I = \frac{V_R}{R}$ (i.e. independent of capacitance)

The energy stored in the capacitor $= \frac{1}{2}CV^2$ where 'V' means the p.d., V_C, across the plates of the capacitor.

EXAM EXAMPLE 36

A student investigates the charging and discharging of a 2200 μF capacitor using the circuit shown.

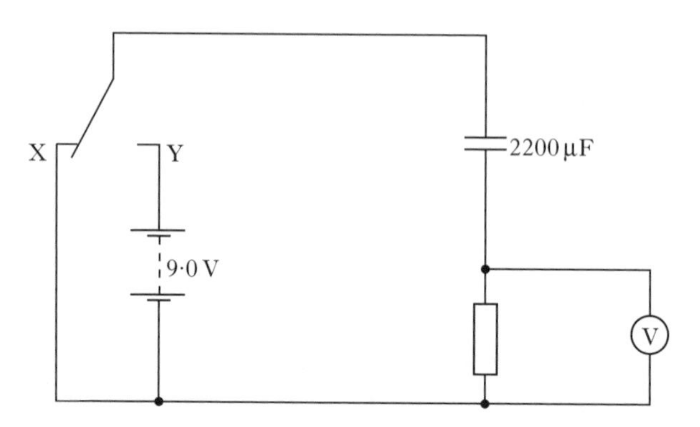

The 9·0 V battery has negligible internal resistance.

Initially the capacitor is uncharged and the switch is at position X.

The switch is then moved to position Y and the capacitor charges fully in 1·5 s.

(*a*) (i) Sketch a graph of the p.d. across the **resistor** against time while the capacitor charges. Appropriate numerical values are required on both axes.

(ii) The resistor is replaced with one of higher resistance.

Explain how this affects the time taken to fully charge the capacitor.

(iii) At one instant during the charging of the capacitor the reading on the voltmeter is 4·0 V.

Calculate the charge stored by the capacitor at this instant.

(*b*) Using the same circuit in a later investigation the resistor has a resistance of 100 kΩ. The switch is in **position Y** and the capacitor is fully charged.

(i) Calculate the maximum energy stored in the capacitor.

(ii) The switch is moved to position X. Calculate the maximum current in the resistor.

Answer:

(a) (i) During charging, the p.d. *across the capacitor* increases from zero to 9·0 V. The p.d. across the resistor therefore decreases from 9·0 V to zero.

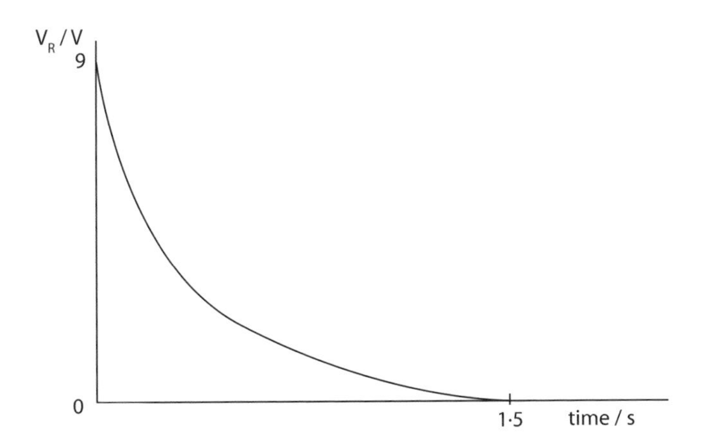

(ii) A greater resistance causes the current at any instant in time to be smaller than before. This means that it takes a *longer time* to transfer the same quantity of charge to fully charge the capacitor.

(iii) $V_R = 4{\cdot}0$ V

so $V_C = V_S - V_R = 9{\cdot}0 - 4{\cdot}0 = 5{\cdot}0$ V

$Q = CV = 2200 \times 10^{-6} \times 5{\cdot}0$
$= 0{\cdot}011$ coulombs ($= 1{\cdot}1 \times 10^{-2}$ C)

(b) (i) When the capacitor is fully charged, in this circuit, $V_C = 9{\cdot}0$ V

$E = \frac{1}{2}CV^2 = \frac{1}{2}\,2200 \times 10^{-6} \times 9{\cdot}0^2$
$= 0{\cdot}0891$ J

(ii) $I_{max} = \frac{V_{max}}{R}$

$= \frac{9{\cdot}0}{100 \times 10^3}$

$= 9{\cdot}0 \times 10^{-5}$ A

- **Poor understanding of the behaviour of a capacitor when connected in a circuit with an a.c. supply**

When a capacitor is connected to a d.c. supply, there is a current in the circuit for only a short time while the capacitor charges. When an alternating supply is used, electrons are made to regularly reverse their direction of flow in the circuit. This causes the capacitor to repeatedly charge one way and then the other. The current does not fall to zero as it does when a capacitor is fully charged using a d.c. source.

- **Lack of knowledge of the relationship between current and frequency in a capacitive circuit**

When an a.c. supply of **low frequency** is connected to a capacitor, electrons flow in one direction for a comparatively long period of time. This means that the capacitor becomes well charged up and the **average current is low.**

When an a.c. supply of **high frequency** is connected to a capacitor, electrons flow in one direction for a comparatively short period of time. This means that the capacitor does not become well charged up and the **average current is high.**

The higher the frequency of the a.c. supply, the greater the average current. In fact, the r.m.s. current is directly proportional to the frequency, i.e.

current-frequency for a *capacitive* circuit

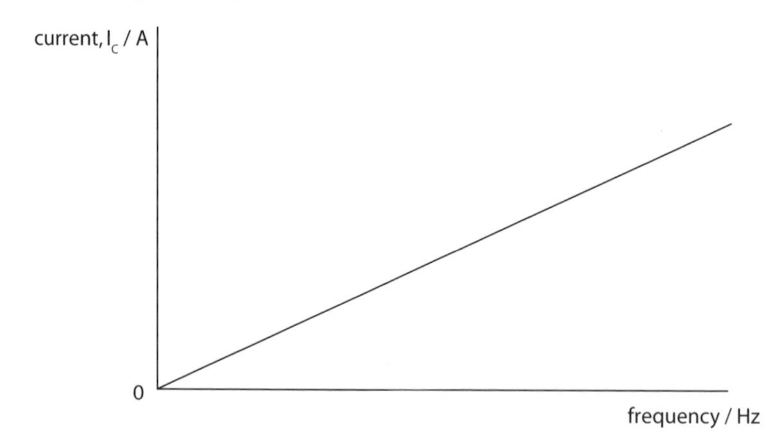

N.B. Do not confuse this with the different behaviour of a *resistor* with an a.c. supply.

EXAM EXAMPLE 37

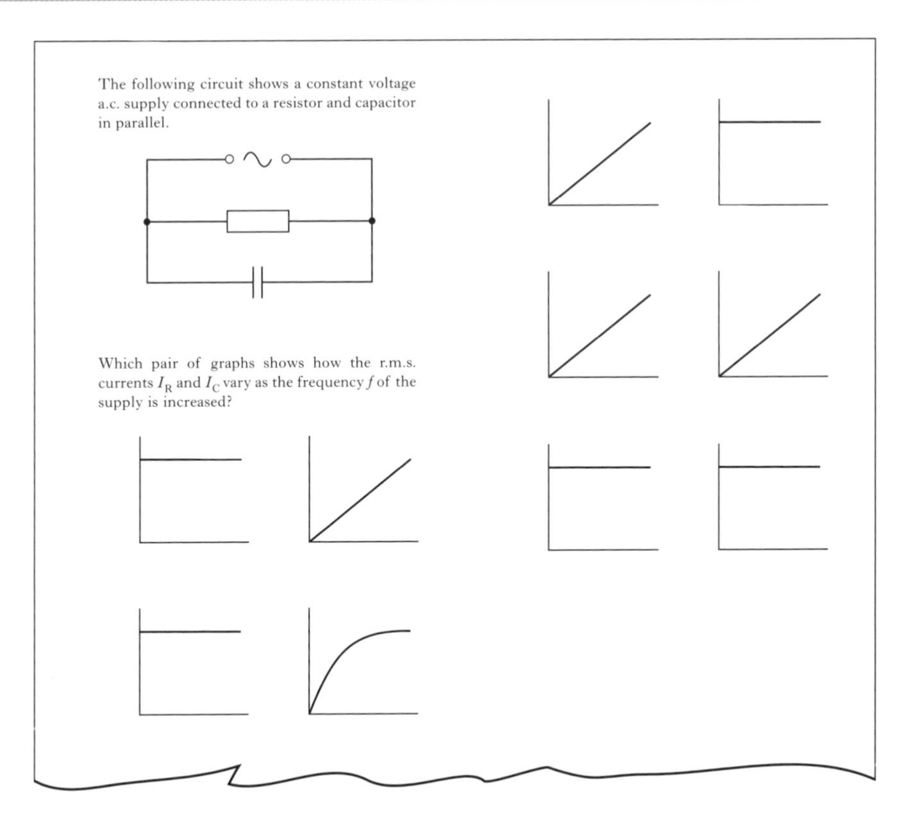

The following circuit shows a constant voltage a.c. supply connected to a resistor and capacitor in parallel.

Which pair of graphs shows how the r.m.s. currents I_R and I_C vary as the frequency f of the supply is increased?

Answer:

I_R is constant at all frequencies.
I_C increases with frequency.

The answer is A.

Analogue Electronics

- **Confusion between input voltages V_1 and V_2 for an op-amp used in differential mode**

There are two input voltages when an op-amp is connected in differential mode. The voltage applied to the inverting input, via input resistor R_1 is labelled V_1. The voltage applied to the non-inverting input, via input resistor R_2 is labelled V_2, i.e.

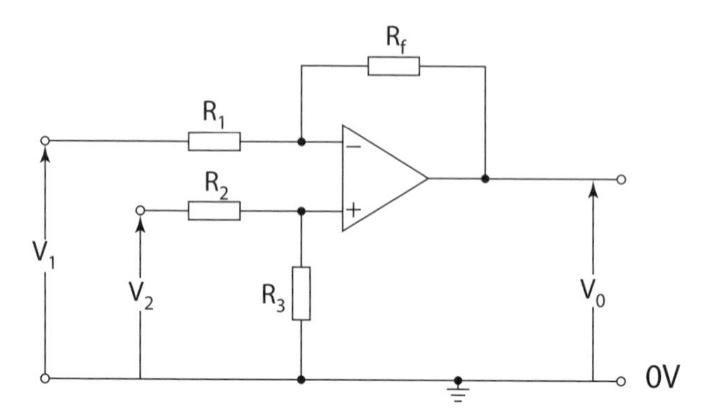

It is **essential** to remember which is V_1 and which is V_2. Substituting them the 'wrong way round' in the formula

$$V_o = (V_2 - V_1) \times \frac{R_f}{R_1}$$

is wrong Physics and loses most of the marks. In English people read from top left, then move right and down – this matches the order of '1' and then '2'.

EXAM EXAMPLE 38

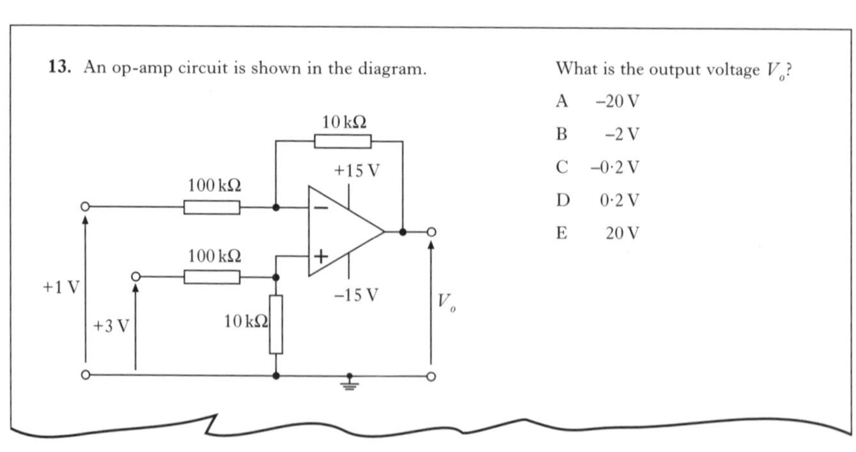

13. An op-amp circuit is shown in the diagram.

What is the output voltage V_o?

A −20 V

B −2 V

C −0·2 V

D 0·2 V

E 20 V

Answer:

$$V_o = (V_2 - V_1) \times \frac{R_f}{R_1} = (3 - 1) \times \frac{10}{100} = +0{\cdot}2 \text{ V}.$$

The answer is D.

If you get mixed up between V_1 and V_2, you are likely to think that *C* is the answer.

- **Incorrect formula used for a given op-amp circuit**

There are two op-amp circuits that need to be *remembered*.

Inverting mode:

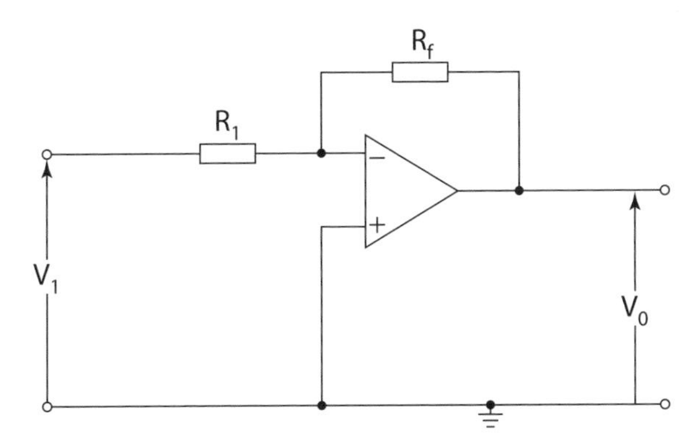

You may decide to remember this circuit by recognising that the non-inverting input is connected directly to the ground line. There is therefore only one input voltage which is replicated in the formula, $\frac{V_o}{V_1} = -\frac{R_f}{R_1}$

Differential mode: see page 74

This circuit has *two* input voltages (V_1 and V_2). The relevant formula is therefore the one which uses *two* input voltages, i.e.

$$V_o = (V_2 - V_1) \times \frac{R_f}{R_1}$$

As mentioned in the previous issue, it is also essential that you remember which input voltage is V_1 and which is V_2.

- **Poor ability to describe/explain the operation of each part of an electronic circuit designed to detect changes in temperature/light and control external devices**

The most common mistake made by candidates is to miss out parts of the circuit or explanation in their description.

Most questions will use a circuit diagram showing a Wheatstone bridge, an op-amp, a transistor and a load device (e.g. a motor). Your answer may need to describe and explain what happens at each stage in this circuit, from how the bridge circuit detects changes in the environment, how the op-amp circuit amplifies the potential difference which is produced, how this can switch on the transistor and how this then operates the load device.

EXAM EXAMPLE 39

(*b*) A student uses this bridge in a circuit to light an LED when the temperature in a greenhouse falls below a certain level.

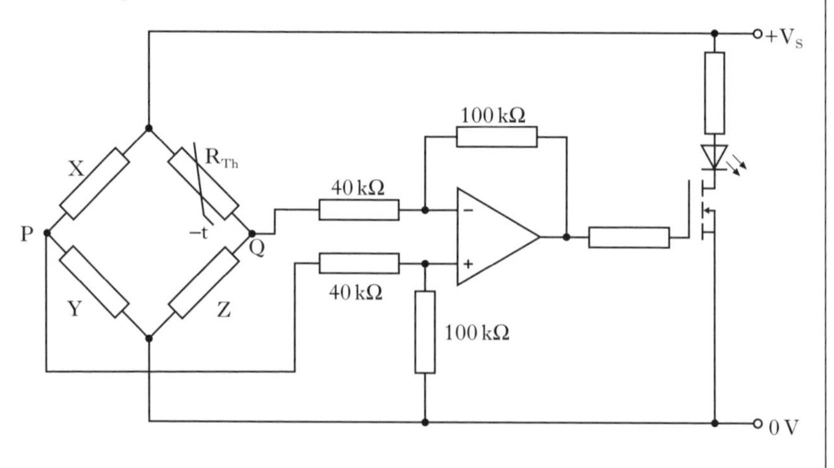

(ii) As the temperature of the thermistor falls, its resistance increases.

Explain how this whole circuit operates to cause the LED to light when the temperature falls.

Answer:

(ii) As the temperature falls the resistance of the thermistor increases. This causes the Wheatstone bridge to go (more) out of balance. This causes a (greater) potential difference between P and Q. This potential difference is amplified by the op-amp circuit. This amplified output voltage is the input voltage to the (MOSFET) transistor. Once this p.d. reaches 2 volts (approx.) the transistor switches on, allowing a current in its output which causes the LED to light.

ADDRESSING AREAS OF WEAKNESS IN RADIATION AND MATTER

Waves

- **Incomplete descriptions of what happens to cause constructive interference and destructive interference**

'Interference' is a wave effect caused by two or more sets of waves *overlapping*. In your answer, you must say that the waves *meet* or 'superpose'. Constructive interference occurs at places where the waves meet in phase with each other. You may describe 'in phase' by saying that 'crests meet crests *and* troughs meet troughs'. Destructive interference occurs at places where the waves meet perfectly out-of-phase with each other (or 'crests meet troughs').

- **Poor understanding of the relationship between path difference and interference patterns**

Path difference is a distance. It is the difference in the distances from the two sources to the point where interference occurs. It is found by subtracting the smaller distance from the larger distance, e.g.

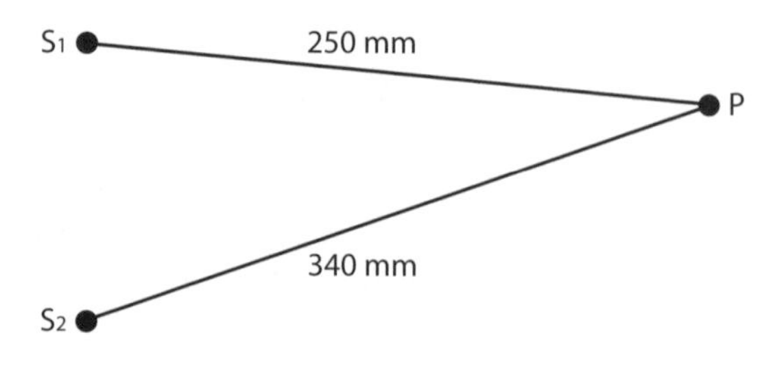

Path difference to P = 340 – 250 = 90 mm

Constructive interference occurs at point P when the path difference is equal to a whole number of wavelengths (because this causes the two sets of waves to meet in phase at P).

Destructive interference occurs at P when the path difference is equal to an odd number of half wavelengths (because this causes the two sets of waves to meet out-of-phase at P).

The relationship between path difference and interference can also be shown as:

	type of interference	order	path difference
	constructive	2	2λ
	destructive	2	$1\frac{1}{2}\lambda$
S_1	constructive	1	λ
	destructive	1	$\frac{1}{2}\lambda$
	constructive	0	0
	destructive	1	$\frac{1}{2}\lambda$
S_2	constructive	1	λ
	destructive	2	$1\frac{1}{2}\lambda$
	constructive	2	2λ

'screen'

EXAM EXAMPLE 40

(*a*) An experiment with microwaves is set up as shown below.

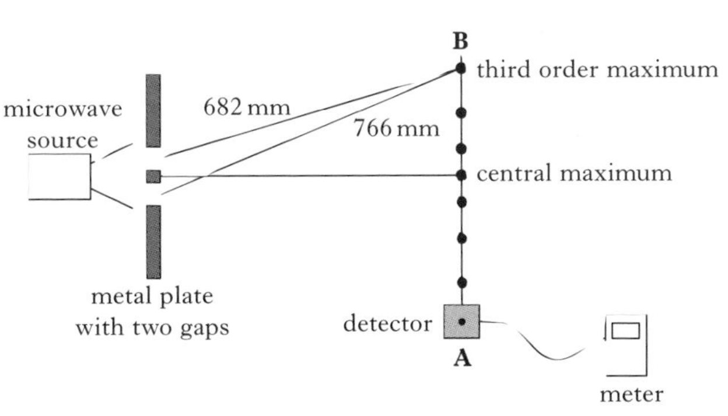

(ii) The measurements of the distance from each gap to a third order maximum are shown. Calculate the wavelength of the microwaves.

Answer:

path difference $= 766 - 682$
$= 84$ mm

Third order maximum means path difference $= 3\lambda$

so $3\lambda = 84$ mm

$\Rightarrow \lambda = \frac{84}{3}$

$\Rightarrow \lambda = 28$ mm

Refraction of Light

- **Poor ability in completing diagrams to show the path taken by a ray of light as it passes from air into and through another medium**

The tools for answering questions like this are:

- knowledge of Snell's law and experience of using it
- knowledge and experience of using some simple geometry

In the Physics data booklet, Snell's law is given as $n = \frac{\sin \theta_1}{\sin \theta_2}$. However, many candidates find it easier to use as $n_1 \sin \theta_1 = n_2 \sin \theta_2$, where the '1' subscript is for values for the medium the ray of light is in initially, and the '2' subscript is for values for the medium the ray of light is travelling to. Angles are measured between the ray and the normal.

You should know that the angles in any triangle add up to 180°. Alternate (or 'z') angles between parallel lines are equal.

EXAM EXAMPLE 41

A physics student investigates what happens when monochromatic light passes through a glass prism or a grating.

(*a*) The apparatus for the first experiment is shown below.

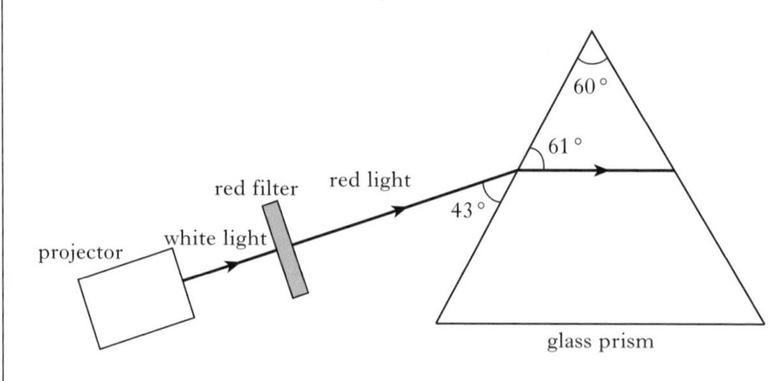

(i) Calculate the refractive index ofthe glass for the red light.

(ii) Sketch a diagram which shows the ray of red light before, during and after passing through the prism. Mark on your diagram the values of all relevant angles.

Answer:

(i) The angles given in the diagram are not between the ray and the normal. In fact, the normal has not been shown and you have to construct (or imagine) it yourself. For the refraction at the left hand side, medium '1' is air and medium '2' is glass.

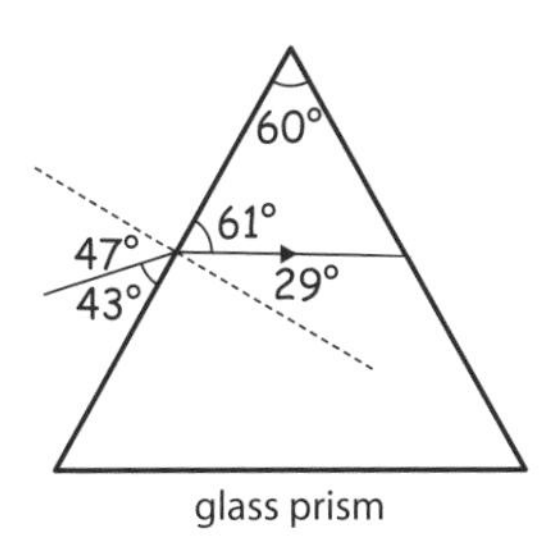

$$n_1 \sin \theta_1 = n_2 \sin \theta_2$$
$$\Rightarrow 1{\cdot}0 \times \sin 47 = n_2 \times \sin 29$$
$$\Rightarrow n_2 = \frac{\sin 47}{\sin 29}$$
$$= 1{\cdot}51$$

(ii) Firstly, use geometry to find the angle of incidence at the right hand face (31°). Then use Snell's law a second time to calculate the angle of refraction outside the right hand face. For the refraction at the right hand side, medium '1' is glass and medium '2' is air.

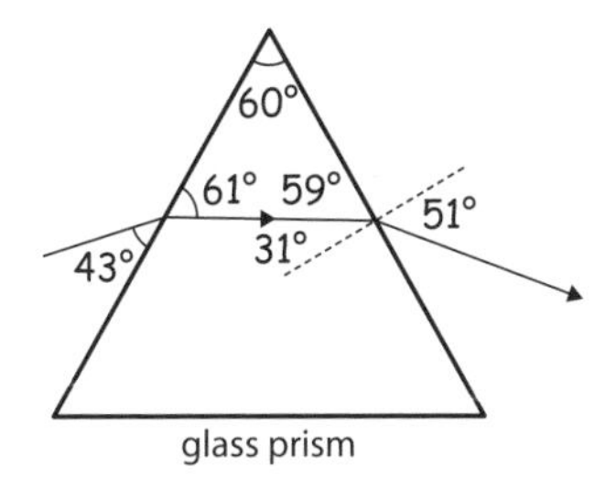

$$n_1 \sin \theta_1 = n_2 \sin \theta_2$$
$$\Rightarrow 1{\cdot}51 \times \sin 31 = 1{\cdot}0 \times \sin \theta_2$$
$$\Rightarrow \sin \theta_2 = 1{\cdot}51 \times \sin 31$$
$$= 1{\cdot}51$$
$$\Rightarrow \theta_2 = 51°$$

- **Lack of knowledge of the link between the refractive index of a medium and the frequency of the light travelling through the medium**

Content Statement 3.2.4 says "State that the refractive index depends on the frequency of the incident light."

The different colours of the visible spectrum have different frequencies. Light at the red end of the spectrum has a lower frequency than light at the violet end of the spectrum. This means that a glass prism has a lower value of refractive index for red light than for, say, blue light. It also means that a ray of blue light will be refracted a greater amount than a ray of red light when sent in along the same initial path.

EXAM EXAMPLE 42

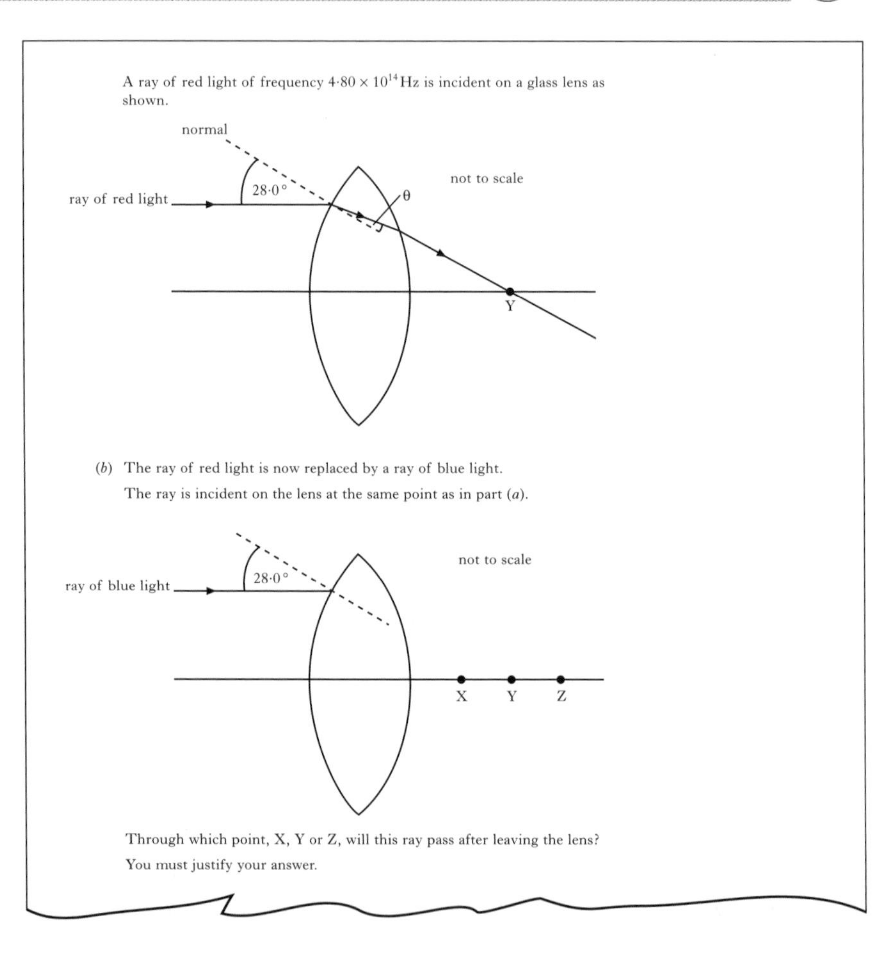

A ray of red light of frequency $4{\cdot}80 \times 10^{14}$ Hz is incident on a glass lens as shown.

(*b*) The ray of red light is now replaced by a ray of blue light.

The ray is incident on the lens at the same point as in part (*a*).

Through which point, X, Y or Z, will this ray pass after leaving the lens?

You must justify your answer.

Answer:

Refractive index depends on the frequency of the light. The glass therefore has a greater refractive index for blue light than for red light. As a result, the ray of blue light is refracted more than the ray of red light. It therefore passes through point X, i.e.

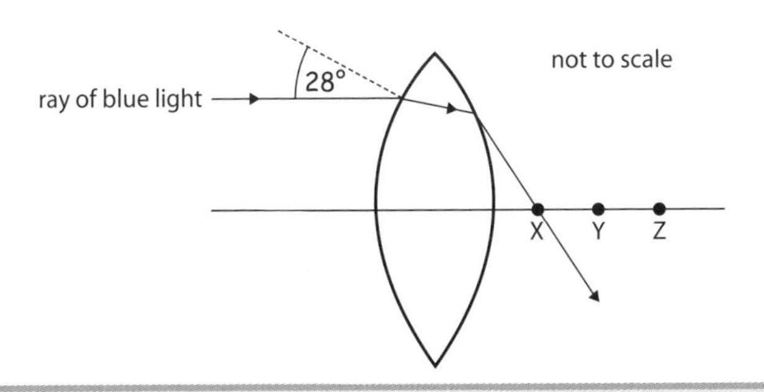

- **Poor explanations of what is meant by total internal reflection**
- **Poor explanations of what is meant by the critical angle, θ_c and the description of a method to measure θ_c**

Total internal reflection can occur when a ray of light is inside a medium of higher refractive index and it is incident on an interface with a medium of lower refractive index (e.g. inside a glass block meeting an interface with air). However, although there is always some of the light internally reflected at this interface, **total** internal reflection only occurs when the angle of incidence is greater than the critical angle, i.e.

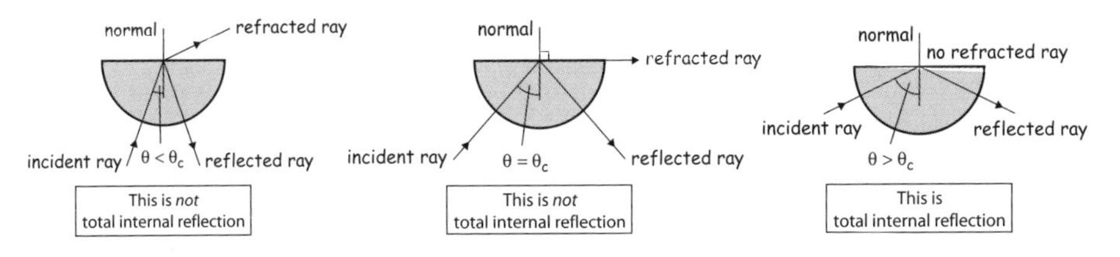

The critical angle, θ_c, is shown in the second diagram. It is the angle at which the refracted ray just emerges along the face of the block. The angle of refraction is 90°. Many candidates wrongly say that it is the angle at which total internal reflection occurs. In fact, total internal reflection only occurs when the angle of incidence is *greater than* the critical angle.

Description of an experiment to measure θ_c:

Use a ray box and semicircular block of glass. A ray of light should be directed through the curved side of the semicircular prism towards the mid-point of the flat face. The angle of incidence should then be increased until the refracted ray just emerges along the straight face. The angle of incidence is then equal to the critical angle and can be measured using a protractor adjusted to find the angle between the incident ray and the normal. Include a diagram (like the second one above) in your description.

Optoelectronics and Semiconductors

- **Poor understanding of the variables and processes involved in the photoelectric effect**

A detailed study of the photoelectric effect requires several pages of a textbook. However, the main points to be known and understood about the photoelectric effect are:

- in some circumstances 'light' can cause a negatively charged metal to discharge
- 'light' can mean infrared, visible or ultraviolet radiation
- the 'light' consists of a stream of photons (bundles of electromagnetic energy)
- one photon collides with one electron and gives that electron all its energy
- the energy carried by each photon is given by $E = hf$ (where 'f' = frequency and 'h' = Planck's constant)
- the minimum energy an electron needs to receive to escape from the metal is called the work function
- an electron is released only if the photon's energy is equal to, or greater than, the work function
- when the photon's energy is greater than the work function, the excess energy becomes kinetic energy of the electron
- increasing the irradiance of the 'light' increases the number of photons per second
- increasing the irradiance of the 'light' increases the number of electrons released per second (only if the photon energy is greater than the work function)
- increasing the irradiance of the 'light' does not increase the maximum kinetic energy of each electron

EXAM EXAMPLE 43

Ultraviolet radiation is incident on a clean zinc plate. Photoelectrons are ejected.

The clean zinc plate is replaced by a different metal which has a lower work function. The same intensity of ultraviolet radiation is incident on this metal.

Compared to the zinc plate, which of the following statements is/are true for the new metal?

I The maximum speed of the photoelectrons is greater.

II The maximum kinetic energy of the photoelectrons is greater.

III There are more photoelectrons ejected per second.

A I only

B II only

C III only

D I and II only

E I, II and III

Answer:

The source of radiation is not changed and so photons have the same energy as before. This means that electrons receive the same energy as before.

A lower work function means that an electron needs less energy to escape from the surface of the metal. This means that electrons have greater kinetic energy than before. Statement II is correct.

Greater kinetic energy means greater speed. Statement I is correct.

The same intensity (now called 'irradiance') means there are the same number of photons per second. Therefore the same number of electrons are released per second. Statement III is incorrect.

The answer is D.

- **Confusion between the terms 'threshold frequency' and 'work function'**

The work function of a metal is the minimum energy an electron needs to receive in order to escape from the surface of a metal. It is measured in joules. The threshold frequency, f_o, is the minimum frequency of 'light' required to cause the emission of electrons (photoemission). It is measured in hertz.

The relationship between these two quantities is

work function = hf_o

- **Inability to explain the effect of increasing irradiance on both the photoelectric current and the maximum kinetic energy of photoelectrons**

Increasing the irradiance of 'light' increases the number of photons per second. This increases the number of electrons released per second and so increases the photoelectric current. However, an electron only receives the energy of one photon. As long as the same frequency of light is used, each photon still has the same energy as before and so there is no change to the kinetic energy of an emitted electron.

EXAM EXAMPLE 44

When light of frequency f is shone on to a certain metal, photoelectrons are ejected with a maximum velocity v and kinetic energy E_k.

Light of the same frequency but twice the irradiance is shone on to the same surface.

Which of the following statements is/are correct?

I Twice as many electrons are ejected per second.

II The speed of the fastest electron is $2v$.

III The kinetic energy of the fastest electron is now $2E_k$.

A I only

B II only

C III only

D I and II only

E I, II and III

Answer:

Twice the irradiance means there are double the number of photons per second. This doubles the number of electrons released per second. Statement I is correct.

Using light of the same frequency means each photon has the same energy as before. This means that each electron receives the same energy as before. Each electron therefore has the same kinetic energy and same speed as before. Statements II and III are incorrect.

The answer is A.

- **Poor understanding of electron transitions in atoms and the absorption or emission of light**

In an atom, electrons go around the nucleus in orbits or energy levels. When an atom absorbs energy, electrons move up from lower energy levels to higher energy levels. When **electrons move down** from higher energy levels to lower levels, the **atom emits energy** (e.g. as a photon of light). The larger the gap between these energy levels, the greater the energy of the emitted photon. This means that when electrons fall through a larger gap between energy levels, light of higher frequency and smaller wavelength is emitted.

EXAM EXAMPLE 45

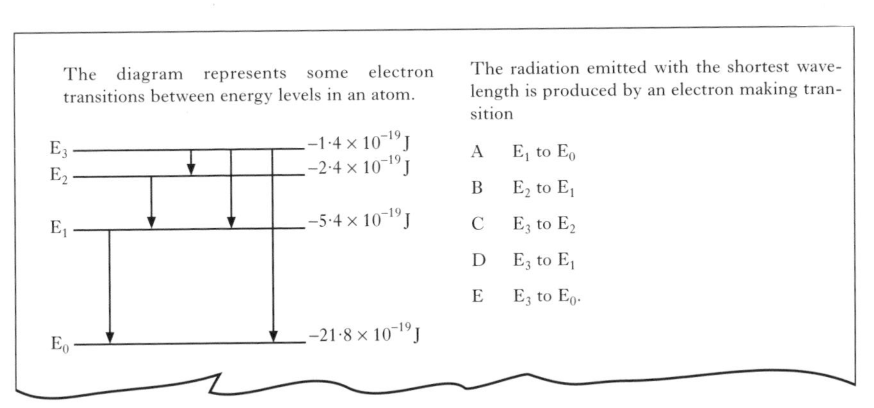

The diagram represents some electron transitions between energy levels in an atom.

E_3 $-1{\cdot}4 \times 10^{-19}$ J
E_2 $-2{\cdot}4 \times 10^{-19}$ J
E_1 $-5{\cdot}4 \times 10^{-19}$ J
E_0 $-21{\cdot}8 \times 10^{-19}$ J

The radiation emitted with the shortest wavelength is produced by an electron making transition

A E_1 to E_0

B E_2 to E_1

C E_3 to E_2

D E_3 to E_1

E E_3 to E_0.

Answer:

Shortest wavelength corresponds to highest frequency.
Highest frequency means greatest energy (because $E = hf$).
Greatest energy means a downward transition through the largest gap, i.e. E_3 to E_0.

The answer is E.

- **Poor understanding of what it means to 'dope' a semiconductor material and the effect this has on conduction**

In a *pure* semiconductor material, such as silicon, all the atoms are the same kind. A pure semiconductor has quite a high resistance, especially at low temperatures, and there is a low current when it is placed in a circuit.

When a small proportion of impurity atoms (such as indium or arsenic) are diffused into the semiconductor, its resistance decreases and it conducts better. Adding these impurity atoms is called 'doping' the semiconductor.

Doping using atoms which have five electrons in their outer shell increases the number of free *electrons* available for conduction. This is an n-type semiconductor. Doping using atoms which have three electrons in their outer shell increases the number of positive *holes* available for conduction. This is a p-type semiconductor.

EXAM EXAMPLE 46

17. A student writes the following statements about p-type semiconductor material.

I Most charge carriers are positive.

II The p-type material has a positive charge.

III Impurity atoms in the material have 3 outer electrons.

Which of these statements is/are true?

A I only

B II only

C I and II only

D I and III only

E I, II and III

Answer:

p-type semiconductors are produced by doping using impurity atoms which have three electrons in their outer shells. Statement III is correct.

This produces gaps in the bonding structure called 'holes', which behave like positive charge carriers. There are now more holes than free electrons. Statement I is correct.

Statement II is incorrect because the added impurity atoms have the same number of positive protons as negative electrons and so are electrically neutral, as was the original semiconductor.

The answer is D.

- **Poor understanding of the processes occurring in an LED to produce light**

An LED is a semiconductor p-n junction diode. When an LED is forward biased, holes in the p-type region and electrons in the n-type region are each pushed towards the junction. At the junction, an electron recombines with a hole and energy is released in the form of a photon of light.

EXAM EXAMPLE 47

18. A p-n junction diode is forward biased.

Positive and negative charge carriers recombine in the junction region. This causes the emission of

A a hole

B an electron

C an electron-hole pair

D a proton

E a photon.

Answer:

The information is describing the behaviour of a light emitting diode (LED). An LED lights due to the emission of photons (one photon being produced for each recombination of an electron and a hole at the junction).

The answer is E.

- **Poor understanding of the processes occurring in a photodiode to produce an e.m.f.**

A photodiode is also a semiconductor p-n junction diode. When photons of light are incident on the junction, electrons are *separated* from holes and an e.m.f. is produced.

- **Lack of ability to identify/draw a photodiode in the photovoltaic or photoconductive mode**

When used in the photovoltaic mode, no external power supply is connected. The e.m.f. produced by the incident photons operates a load (e.g. a motor) directly. In the photoconductive mode a photodiode is connected in series with a resistor and in reverse bias with an external power supply.

EXAM EXAMPLE 48

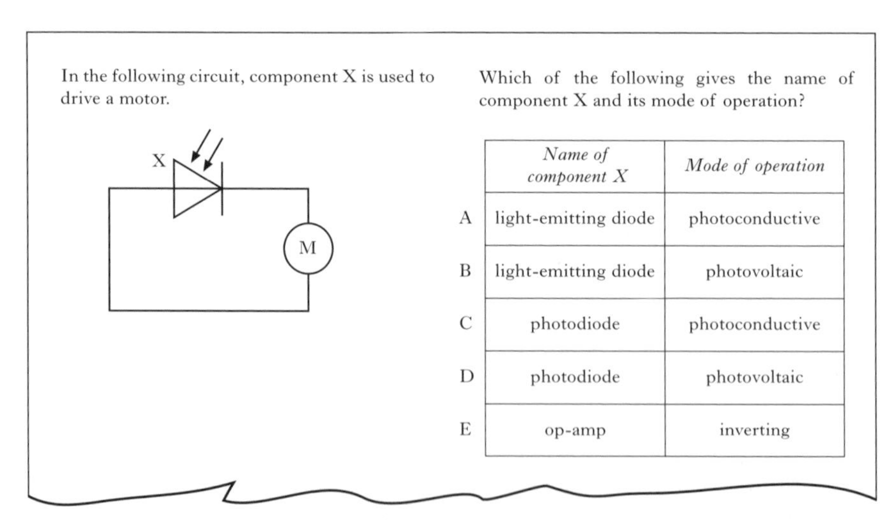

In the following circuit, component X is used to drive a motor.

Which of the following gives the name of component X and its mode of operation?

	Name of component X	*Mode of operation*
A	light-emitting diode	photoconductive
B	light-emitting diode	photovoltaic
C	photodiode	photoconductive
D	photodiode	photovoltaic
E	op-amp	inverting

Answer:

The symbol has two arrows pointing towards the diode. This represents light incident *onto* the diode. It is therefore a photodiode. (Two arrows pointing *away* would indicate the *emission* of light, which would be the symbol for an LED.)

There is no external power supply. The photodiode is powering the motor by the e.m.f. generated by the incoming photons. This photodiode is being used in the photovoltaic mode.

The answer is D.

The following circuit shows a photodiode connected in photoconductive mode.

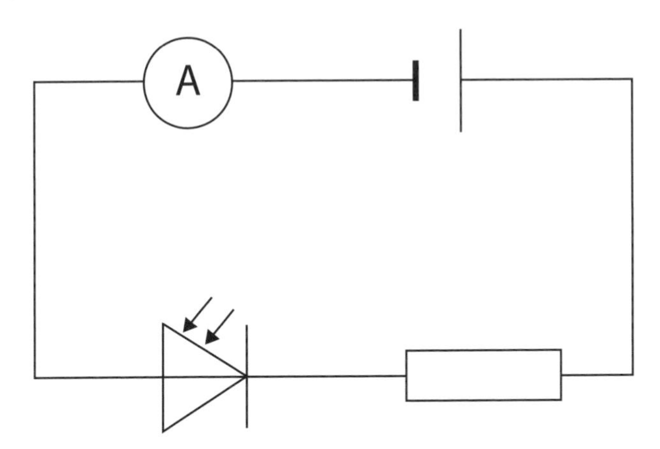

- an external power supply is connected to the photodiode
- the polarity of the external power supply makes the photodiode reverse biased

Nuclear Reactions

- **Poor ability to describe Rutherford's experiment and how its results led to our model of the atom**

In Rutherford's experiment a beam of alpha particles was directed at a very thin gold foil. A detector was moved round the foil to discover where the alpha particles went after colliding with the gold. Most of the alpha particles carried on straight through the foil, showing that an atom is mainly empty space. A very small number of alpha particles were deflected back the way they had come, showing that atoms have a very small, positive centre where most of the mass is concentrated (the nucleus).

EXAM EXAMPLE 49

The classical experiment on the scattering of alpha particles from a thin gold foil suggested that

A positive charges were evenly distributed throughout the atom

B atomic nuclei were very small and positively charged

C neutrons existed in the nucleus

D alpha particles were helium nuclei

E alpha particles were hydrogen nuclei.

Answer:

A – an older model of the structure of an atom

B – **the correct answer**

C – a true statement, but not connected with this experiment

D – a true statement, but not connected with this experiment

E – an incorrect statement

- **Confusion between the processes of nuclear fission and nuclear fusion**

Nuclear fission is the process of a large nucleus breaking down into smaller nuclei. There is a loss of total mass and energy is produced, according to $E = mc^2$.

Nuclear fusion is the process of two small nuclei joining together to form a larger nucleus. There is a loss of total mass and energy is produced, according to $E = mc^2$.

- **Rounding values of masses before the loss of mass is calculated**

In both nuclear fission and nuclear fusion there is a very small loss in mass. This lost mass becomes energy according to Einstein's equation, $E = mc^2$. However, the loss in mass is so small that significant inaccuracies are introduced if figures are rounded before calculating its value. For this reason, in questions on fission and fusion, masses are often given to six significant figures. Do not round answers to calculations until the final value for the energy released.

Dosimetry and Safety

- **Inability to define the 'activity' of a radioactive source**

Content Statement 3.5.1 says 'State that the activity of a radioactive source is the number of decays per second and is measured in becquerels (Bq), where one becquerel is one decay per second.'

EXAMPLE 50

What is meant by an *activity of* 8kBq?

Answer:

An activity of 8 kBq means that 8 000 atoms decay each second.

- **Confusion between the terms 'absorbed dose', 'equivalent dose' and 'equivalent dose rate'**

Absorbed dose, *D*, is the energy of the radiation absorbed per kilogram of tissue.

$$D = \frac{E}{m}$$

Equivalent dose, *H*, is the absorbed dose multiplied by the radiation weighting factor.

$$H = D\,w_R$$

Equivalent dose rate, $\dot{H}$ is the equivalent dose divided by time.

$$\dot{H} = \frac{H}{t}$$

ADDRESSING GENERAL AREAS OF WEAKNESS

Units, Prefixes and Scientific Notation

- **Some SI units not known well enough**

You must learn the units of all the quantities in the Higher Physics course. It is useful to go through your notes or textbook and make your own list. However, you may wish to download the SQA's list of Quantities, Symbols and Units.

- **Rounding the answers to intermediate calculations, leading to inaccuracies in the final answer**

Some questions require you to carry out a 'double' calculation. The answer to the first calculation is used in the second calculation. Rounding the first answer too much can cause inaccuracy in the answer to the second calculation.

EXAM EXAMPLE 51

(*a*) A ray of red light of frequency $4{\cdot}80 \times 10^{14}$ Hz is incident on a glass lens as shown.

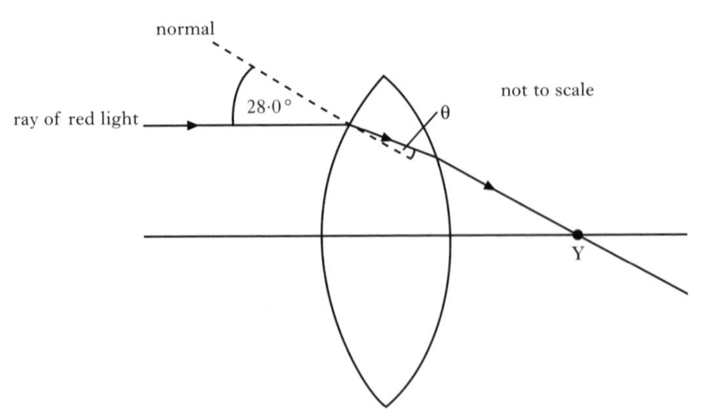

The ray passes through point Y after leaving the lens.

The refractive index of the glass is 1·61 for this red light.

(ii) Calculate the wavelength of this light inside the lens.

Correct answer:

Wavelength in air, $\lambda_{air} = \frac{v_{air}}{f} = \frac{3{\cdot}00 \times 10^{8}}{4{\cdot}80 \times 10^{14}} = 6{\cdot}25 \times 10^{-7}$ (m)

Wavelength in glass, $\lambda_{glass} = \frac{\lambda_{air}}{n} = \frac{6{\cdot}25 \times 10^{-7}}{1{\cdot}61} = 3{\cdot}88 \times 10^{-7}$ m

Incorrect answer:

Wavelength in air, $\lambda_{air} = \frac{v_{air}}{f} = \frac{3{\cdot}00 \times 10^{8}}{4{\cdot}80 \times 10^{14}} = 6{\cdot}25 \times 10^{-7} = 6 \times 10^{-7}$ (m)

Wavelength in glass, $\lambda_{glass} = \frac{\lambda_{air}}{n} = \frac{6 \times 10^{-7}}{1{\cdot}61} = 3{\cdot}73 \times 10^{-7}$ m

Here the final answer, although given to three significant figures, is inaccurate because the calculation for the wavelength in air was inappropriately rounded to one significant figure.

- **Too many or too few significant figures given in the final answer**

Final answers should be rounded 'to an appropriate number of significant figures'. To decide what is 'appropriate' in a given question you need to consider the number of figures in the given data. The final answer should be rounded to the same number of significant figures as the least precise data item.

Inappropriate rounding can result in a marking penalty in the examination. In practice, most final answers in Higher Physics can be rounded to two or three significant figures without penalty.

- **Incorrect rounding of final answers**

For example a figure of '26·57', rounded to three significant figures should be '26·6', not '26·5'

EXAMPLE

Suppose the data in a particular question is given to three significant figures. Your calculator displays the answer to the calculation as $4{\cdot}738709677 \times 10^{14}$. The third significant figure in this answer is the '3'. However, because the fourth figure is '8', the '3' is rounded up and the answer should be written as $4{\cdot}74 \times 10^{14}$. Do not round up when the next figure is less than '5'.

- **Some prefixes not known well enough**

For this course the prefixes to be known are:

Prefix name	Prefix symbol	Power of ten
pico	p	$\times 10^{-12}$
nano	n	$\times 10^{-9}$
micro	μ	$\times 10^{-6}$
milli	m	$\times 10^{-3}$
kilo	k	$\times 10^{3}$
Mega	M	$\times 10^{6}$
Giga	G	$\times 10^{9}$

These prefixes are not listed on the examination paper or given in the data booklet. The only way of knowing them for the examination is to learn them.

Uncertainties

- **Inability to change from absolute uncertainty to percentage uncertainty and vice versa**

EXAMPLE

A student plans to drop a ball from a height. She measures the height as 2.50 metres. She estimates that the uncertainty in this measurement is ±2 centimetres (±0·02 m) – this means that she believes the value of the height lies between 2·48 and 2·52 metres.

She should record her result as 2·50 ±0·02 m.

The *absolute* uncertainty in her measurement is the value of the imprecision, i.e. ±0·02 m.

The *fractional* uncertainty in her measurement is the absolute uncertainty divided by the value of the measurement, i.e. $\frac{0{\cdot}02}{2{\cdot}50} = 0{\cdot}008$.

The *percentage* uncertainty in her measurement is the fractional uncertainty multiplied by 100, i.e. $0{\cdot}008 \times 100 = 0{\cdot}8\ \%$.

- **Poor ability to express the final numerical result of an experiment in the form: final value ± uncertainty**

For Higher Physics you need to know that the best estimate of the percentage uncertainty in a final answer is equal to the bigger (or biggest) percentage uncertainty in the measurements used to calculate that answer. You must also be able to calculate an absolute uncertainty from a percentage uncertainty.

EXAMPLE

A student is given the following measurements:

Current = $0{\cdot}025 \pm 0{\cdot}001$ A
Voltage = $12{\cdot}0 \pm 0{\cdot}25$ V
The student is asked to calculate the resistance and give its *absolute* uncertainty.

Answer:

Resistance, $R = \frac{V}{I} = \frac{12}{0{\cdot}025} = 480\ \Omega$.

The absolute uncertainty in the current is $\pm 0{\cdot}001$ A.

The fractional uncertainty in the current is $\frac{0{\cdot}001}{0{\cdot}025} = 0{\cdot}04$.

The percentage uncertainty in the current is 4% (Fractional uncertainty × 100).
The absolute uncertainty in the voltage is $\pm 0{\cdot}25$ V.

The fractional uncertainty in the voltage is $\frac{0{\cdot}25}{12} = 0{\cdot}0208$

The percentage uncertainty in the voltage is 2%.
The uncertainty in the resistance is therefore ±4% (the bigger value).
4% of 480 = $480 \times \frac{4}{100} = 19$, which rounds to 20.

The resistance is therefore 480 ±20 Ω.

Precision of language

- **Lack of precision in the choice of words**

Physicists are often accused of being 'nit pickers'. Whether this is a fair description or not, it is certainly true that a high level of precision is demanded to ensure good communication. In Higher Physics, candidates are expected to reflect that in their answers.

Here are some examples of poor use of language along with suggested improvements.

Candidate's answer	Improved language use	Comments
Acceleration speeds up when mass decreases	Acceleration increases when mass decreases	Only an *object* can be said to speed up (when its velocity changes)
Gas pressure is due to molecules colliding	The pressure a gas exerts is due to molecules colliding with the walls of the container	Molecules also collide with each other, but this does not cause a force on the walls
When the temperature of a gas is increased, the pressure increases because molecules collide more	When the temperature of a gas is increased, the pressure increases because molecules collide with the walls of the container more frequently and harder	Increasing the temperature changes *two* factors which each cause an increase in the force exerted on the walls of the container.
The time taken for the capacitor to charge is faster	The time taken for the capacitor to charge is shorter	Times are longer or shorter. Only speeds are faster or slower.
Current is increased when a smaller resistor is used *Current is increased when a lower resistor is used*	Current is increased when a smaller resistance is used OR Current is increased when a smaller value of resistor is used	'Smaller' is to do with physical size. It is possible to have a tiny size of resistor which has a very high value of resistance. 'Lower' is a description of height
Alternating current means that electrons can flow both ways round a circuit	Alternating current means that electrons regularly reverse their direction of flow in a circuit	Saying that they *can* – flow both ways does not mean that they *do* change direction.

- **Technical language not used appropriately**
- **Specialised words spelt wrongly**

Make sure you do not lose marks by making any of the following mistakes.

Word/terminology	Common mistakes
weight	*gravity*
upthrust	*upforce*
reflection	*refraction* (this term does exist, but means something different)
diffraction	*Defraction* (this word does not exist)
threshold	*freshold* (this word does not exist)
threshold frequency	*work function* (this term does exist, but means something different)
silicon	*silicone*
germanium	*geranium*
fission	*fusion, fussion*
fusion	*fission, fussion*
half-value thickness	*half life* (this term does exist, but means something different)

Graphs

- **Failure to label the origin with a zero**

Many graphs in Physics are used to indicate a relationship between quantities, for example that the pressure of a gas is directly proportional to absolute temperature. A graph can prove such a relationship when it is a straight, diagonal line through the origin. The intersection of the axes must therefore be shown as the point where both the quantities have their zero value. The usual way to do this is to write a zero ('0') at this point. Failure to do this usually results in a loss of marks.

- **Failure to fully label the axes of sketch graphs**

The axes of all graphs (whether precisely drawn on graph paper or just sketched on plain paper) should have each axis labelled with both the name of the quantity and its units. Any relevant values should also be marked on the axis and dotted reference lines drawn from them to the graph line.

- **Incorrect reading of scales of graphs in questions**

Take time to work out the value of one division along each axis. Double check your answer by making sure that you agree with the numbers given further along the axes.

- **Sketch graphs too 'rough'**

Make sure you use a ruler to draw the axes and the graph line (if it is straight). Take care to show whether your graph line intersects an axis (e.g. initial charging current with a capacitor) or whether it does not touch the axis (e.g. the activity of a radioactive source never reaches zero). Draw dotted reference lines to indicate any important values.

Discussions/explanations based on formulas

- **Failure to quote the relevant formula**

When giving an explanation, check whether a formula is associated with it (for example, by looking through the Physics data booklet). Write the formula into your answer.

- **Failure to state whether each quantity increases, decreases or stays constant**

When giving an explanation in which you have included a formula, make sure you describe what happens to each of the quantities – even when their values do not change.

Miscellaneous

- **Lack of familiarity with the list of Content Statements for the course**

As part of your regular study and revision, make sure you study each of the Content Statements. These can be downloaded from SQA's website (as part of the Arrangements document for Higher Physics). You can also use these as a checklist to ensure you have covered the course.

- **Lack of detail in answers to 'show' questions**

When asked to "show" that a value is true, it is essential that you quote any relevant formula, substitute appropriate values and show the details of calculations. Missing out any of these stages means that a marker cannot be sure whether you have worked 'backwards' from the answer.

- **Numerical analyses not presented in a clear and structured way**

The answers given by candidates sometimes only show various numbers across a page without any obvious reason for choosing them or any connection between them. This can make it very difficult for a marker to identify any correct Physics meriting the award of marks. Make sure you structure your answers going down the page as follows:

i quote the relevant formula

ii substitute values into the appropriate places

iii work out the unknown value

iv give units after the final numerical answer

- **Failing to use the same labelling for an answer as given for the question**

Obviously all answers need to be numbered. Do not miss out this numbering or make up your own numbering system – this can cause marks to be lost.

- **Giving more than one answer to a question**

It should be obvious that a marker cannot be expected to choose between two different answers. Whenever you make more than one attempt at an answer, make sure you make a final decision and *clearly delete* the one you wish to be ignored by a marker.

7 Analysing the Quality of Answers

An important skill which helps produce top quality answers is to be able to recognise strengths and weaknesses in example answers and then to reproduce only the strengths in your own answers.

This section of the book shows questions which have caused difficulties for candidates in previous SQA examinations. For each of these questions, example answers are given. Being typical of answers given by candidates, some are wrong or weak, some are good and some are excellent.

It can be useful just to read through the answers and the analysis of them. However, the most effective way to use the material in this section is to carry out the following series of tasks:

- consider one question at a time
- read that question carefully, as if you were sitting in the examination room
- write out your own answer to the question
- read and compare the different example answers
- try to recognise which are the better/best answers
- identify the strengths which make the better answer(s) superior
- identify what is wrong/missing in the weaker answer(s)
- compare your ideas with the 'correct' analysis of the answers
- memorise the points which make the excellent answer(s) so good.

QUESTION 1

Competitors are racing remote control cars. The cars have to be driven over a precise route between checkpoints.

Each car is to travel from checkpoint A to checkpoint B by following these instructions.

"Drive 150 m due North, then drive 250 m on a bearing of 60° East of North (060)."

Car X takes 1 minute 6 seconds to follow these instructions exactly.

(*a*) By scale drawing or otherwise, find the displacement of checkpoint B from checkpoint A.

(*b*) Calculate the average velocity of car X from checkpoint A to checkpoint B.

(*c*) Car Y leaves A at the same time as car X.

Car Y follows exactly the same route at an average speed of $6{\cdot}5\,\mathrm{m\,s^{-1}}$.

Which car arrives first at checkpoint B?

Justify your answer with a calculation.

(*d*) State the displacement of checkpoint A from checkpoint B.

Candidate A's answer:

(a)

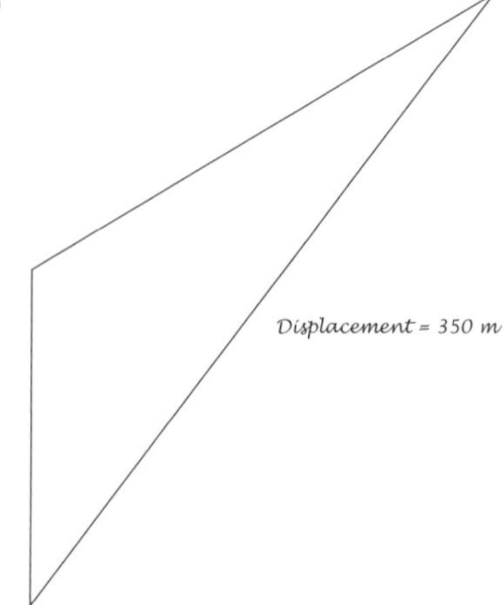

(b) $v = \frac{d}{t} = \frac{400}{66} = 6{\cdot}06\ m\,s^{-1}$

(c) car Y arrives first because it is faster

(d) 350 m

Candidate B's answer:

(a)

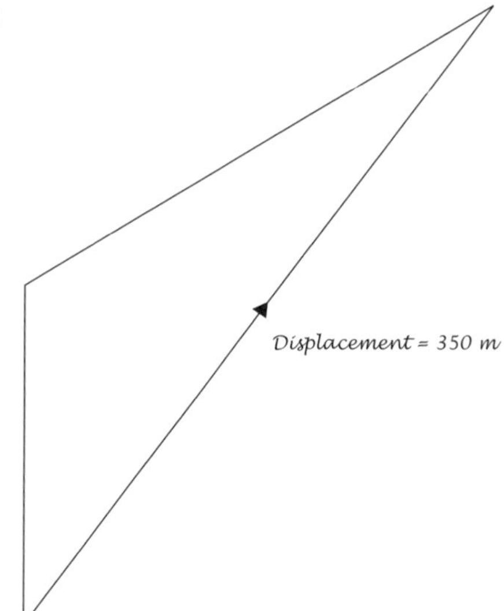

(b) $v = \frac{d}{t} = \frac{350}{66} = 5.3$

(c) car Y arrives first

(d) 350 m

Candidate C's answer:

(a)

scale: 1 cm : 50 m

B

60°

Displacement = 350 m bearing 038°

38°

A

(b) 1 minute 6 seconds = 60 + 6 = 66 seconds.

$v = \frac{s}{t} = \frac{350}{66} = 5.3\ m\,s^{-1}$ bearing 038°

(c) $time_Y = \frac{distance}{speed} = \frac{400}{6.5} = 61.5$ seconds.

This time is less than the 66 s taken by car X, so car Y arrives first.

(d) Displacement is the same size, but in the opposite direction. 038 + 180 = 218
Displacement of A from B = 350 m bearing 218°

Analysis of example answers

Candidate A: This answer is weak.

Part	Strengths	Weaknesses
(a)	• the value of the final displacement is correct	• no scale is given for the diagram • the origin (checkpoint A) has not been labelled • checkpoint B has not been labelled • no arrows have been drawn on any of the vectors • the direction of the total displacement has not been quoted
(b)	• the formula may be appropriate	• the value of '400' is the total distance travelled – to calculate average velocity the calculation must be of *total displacement/time*
(c)	• the conclusion is correct • a reason has been given	• the answer does not show a calculation (as asked for in the question)
(d)	• the *value* of the displacement is correct	• no direction has been given (a vector is not fully defined until its direction is quoted)

Candidate B:

Part	Strengths	Weaknesses
(a)	• the value of the final displacement is correct • an arrow has been drawn on the vector representing the final displacement	• no scale is given for the diagram • the origin (checkpoint A) has not been labelled • checkpoint B has not been labelled • no arrows have been drawn on the initial vectors • the direction of the total displacement has not been quoted
(b)	• the value of the average velocity has been correctly calculated from total displacement/time	• the units for velocity are missing ($m s^{-1}$) • no direction has been quoted for velocity (which is a vector quantity)
(c)	• the conclusion is correct	• no reason has been given • the answer does not show a calculation (as asked for in the question)
(d)	• the *value* of the displacement is correct	• no direction has been given

Candidate C: This is an excellent answer.

All parts are answered correctly *and* they are clearly explained with good detail given.

QUESTION 2

A "giant catapult" is part of a fairground ride.

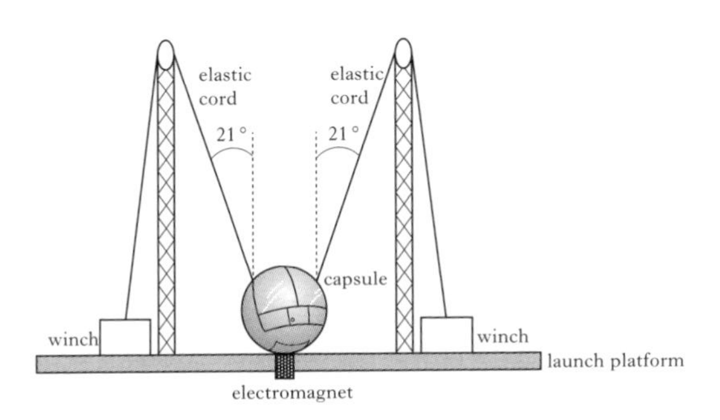

Two people are strapped into a capsule. The capsule and the occupants have a combined mass of 236 kg.

The capsule is held stationary by an electromagnet while the tension in the elastic cords is increased using the winches.

The mass of the elastic cords and the effects of air resistance can be ignored.

(*a*) When the tension in each cord reaches $4{\cdot}5 \times 10^3$ N the electromagnet is switched off and the capsule and occupants are propelled vertically upwards.

(i) Calculate the vertical component of the force exerted by **each** cord just before the capsule is released.

(ii) Calculate the initial acceleration of the capsule.

(iii) Explain why the acceleration of the capsule decreases as it rises.

(*b*) Throughout the ride the occupants remain upright in the capsule.

A short time after release the occupants feel no force between themselves and the seats.

Explain why this happens.

Candidate A's answer:

(a) *(i)* $F_v = F\cos\theta = 4.5 \times 10^3 \times \cos 21 = 4201.112 = 4200$ N

(ii) *Total upward force* $= 2 \times 4200 = 8400$ N

Downward force = weight of capsule and occupants $= mg = 236 \times 9.8 = 2312.8 = 2313$ N

Resultant force $= 8400 - 2300 = 6100$ N *upwards*

acceleration $= F/m = 6100/236 = 25.847\ ms^{-2} = 26\ ms^{-2}$ *upwards.*

(iii) Gravity has now had time to pull the capsule back down.

(b) When the cords are no longer stretched, they exert no upwards force on the capsule and so both the occupants and the capsule have the same downward acceleration of $9{\cdot}8\ m\ s^{-2}$.

Candidate B's answer:

(a) (i) $F_v = F \sin \theta = 4.5 \times 10^3 \times \sin 21 = 1612.6558 = 1613$ N

(ii) Total force $= 2 \times 1613 = 3226$ N

acceleration $= F/m = 3226/236 = 13.7$ ms^{-2} upwards

(iii) The capsule is now further away from the Earth.

(b) The occupants are no longer accelerating.

Candidate C's answer:

(a) (i) $F_v = F \cos \theta = 4.5 \times 10^3 \times \cos 21 = 4201.112$ N

(ii) Total upward force $= 2 \times 4201.112 = 8402.224$ N

Downward force = weight of capsule and occupants $= mg = 236 \times 9.8 = 2312.8$

Resultant force $= 8402.224 - 2312.8 = 6089.424$ N upwards

acceleration $= F/m = 6089.424/236 = 25.8026$ ms^{-2} upwards

(iii) As the capsule rises, the angle of the cord to the vertical increases. The component of the upward force therefore decreases.

The unbalanced force, F, on the capsule decreases and so 'a' decreases (because $a = F/m$).

(b) The force between the seats and the occupants is equal to the weight of the occupants.

Analysis of example answers

Candidate A's answer:

Part	Strengths	Weaknesses
(a)(i)	• an appropriate formula has been used correctly • the answer has been appropriately rounded to two significant figures	
(a)(ii)	• this part has been answered perfectly	
(a)(iii)		• gravity is always pulling the capsule down, even when it had its initial high acceleration upwards – this candidate seems to be confused between 'acceleration' and 'velocity'
(b)	• this part has been answered perfectly	

Candidate B's answer:

Part	Strengths	Weaknesses
(a)(i)		• the formula is incorrect – vertical components are only found using the 'sine' formula when the angle is given to the horizontal
(a)(ii)	• the upward force from the cords is found by doubling the answer to part (a)(i) • $F = ma$ is an appropriate formula	• the downward force (= weight) has not been calculated or taken into account to find the unbalanced force, F
(a)(iii)		• there is negligible change to the gravitational field strength of the Earth in the height reached by the capsule
(b)		• the occupants are accelerating at all stages of the ride – accelerating upwards while the force from the cords is greater than the weight – and then accelerating downwards (even when still moving upwards)

Candidate C's answer:

Part	Strengths	Weaknesses
(a)(i)	• an appropriate formula has been used correctly	• there are too many significant figures in the answer (the data is to 2 or 3 figs.)
(a)(ii)	• the Physics of calculating the forces and finding the resultant has been done well	• there are too many significant figures in the final answer (should be 2 or 3 figs.)

Continued

(a)(iii)	• this analysis of the forces acting and the link between resultant force and acceleration is explained perfectly	
(b)		• if this were true, the force would not be zero (as their weight is never zero) {Note that this force = weight, only when the capsule's speed is constant}

The 'perfect' answer would be:

(a)(i) Answer A, (a)(ii) Answer A, (a)(iii) Answer C, (b) Answer A.

QUESTION 3

A crane barge is used to place part of an oil well, called a manifold, on the seabed.

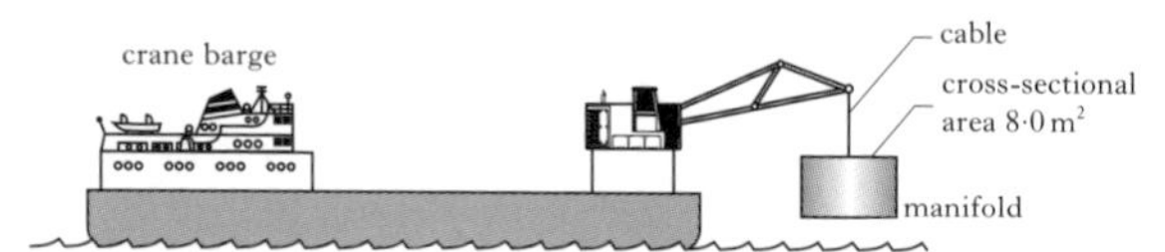

The manifold is a cylinder of uniform cross-sectional area $8{\cdot}0\,m^2$ and mass $5{\cdot}0 \times 10^4$ kg. The mass of the cable may be ignored.

(*a*) Calculate the tension in the cable when the manifold is held stationary above the surface of the water.

(*b*) The manifold is lowered into the water and then held stationary just below the surface as shown.

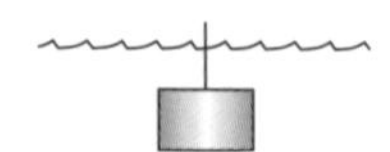

(i) Draw a sketch showing all the forces acting vertically on the manifold. Name each of these forces.

(ii) The tension in the cable is now $2{\cdot}5 \times 10^5$ N.

Show that the difference in pressure between the top and bottom surfaces of the manifold is $3{\cdot}0 \times 10^4$ Pa.

(*c*) The manifold is now lowered to a greater depth.

What effect does this have on the difference in pressure between the top and bottom surfaces of the manifold?

You must justify your answer.

Candidate A's answer:

(a) *tension = weight = mg =* $5.0 \times 10^4 \times 10 = 500\ 000$

(b) (i)

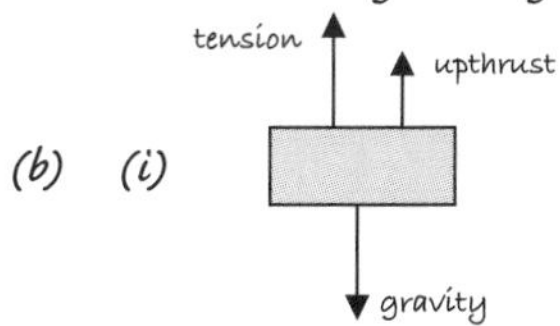

(b) (ii) $P = F/A = 2.5 \times 10^5/8 = 3.125 \times 10^4 = 3 \times 10^4$ Pa

(c) *greater, because* $P = \rho g h$ *and h is greater.*

Candidate B's answer:

(a) 490000 N

(b) (i)

T
u
W

(b) (ii) *upthrust + tension = weight*

Upthrust = weight – tension = $4.9 \times 10^5 - 2.5 \times 10^5 = 2.4 \times 10^5$ N

(c) *In a fluid* $P = \rho g h$ *Both* ρ *and g are constant.*
Although there is an increase in the pressure on the top and the pressure on the bottom of the manifold, they both increase by the same amount and so the difference between them remains the same.

Candidate C's answer:

(a) *tension = weight of manifold = mg =* $5.0 \times 10^4 \times 9.8 = 490\ 000$ N

(b)(ii)

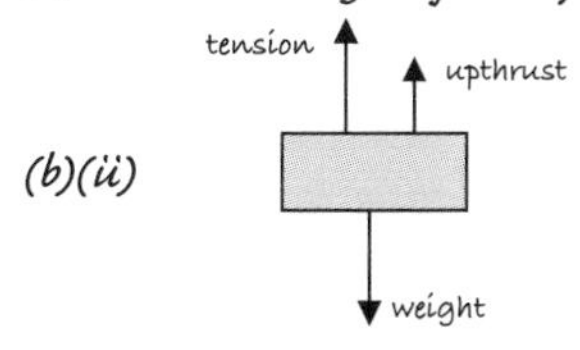

(b)(ii) $\Delta P = \Delta F/A$

$= (4.9 \times 10^5 - 2.5 \times 10^5)/8$

$= 2.4 \times 10^5/8$

$= 3.0 \times 10^4$ Pa

(c) *This has no effect on the pressure difference.*

Analysis of example answers

Candidate A's answer:

Part	Strengths	Weaknesses
(a)	• correct formula	• 'g' should be '9.8', not '10' in Higher Physics • units are missing after the final value
(b)(i)	• tension and upthrust are correctly named and directions shown	• the name of the force due to gravity is weight
(b)(ii)	• the correct formula for pressure is quoted	• the value of tension has been used for 'F' – it should be the upthrust force which is related to the pressure difference
(c)	• relevant formula quoted	• the conclusion is wrong because the question is asking about the pressure difference, not the absolute value of the pressure

Candidate B's answer:

Part	Strengths	Weaknesses
(a)	• the correct answer	• no formula or working has been shown – the candidate could have lost all the marks if an error had been made
(b)(i)		• the question asks for the forces to be named. The labels 'T', 'U' and 'W' are not names – the question has not been answered.
(b)(ii)	• a correct answer, clearly set out and fully explained	
(c)	• an excellent answer, fully and clearly explained	

Candidate C's answer:

Part	Strengths	Weaknesses
(a)	• a correct and clearly explained answer	
(b)(i)	• the correct answer	
(b)(ii)	• the correct answer	• not quite as clearly explained as Candidate B's answer
(c)	• the correct conclusion	• no justification/explanation has been given. The question said that a justification must be given. This answer can receive no marks.

The 'perfect' answer would be:

(a) Answer C, (b)(i) Answer C, (b)(ii) Answer B, (c) Answer B.

QUESTION 4

A cylinder of compressed oxygen gas is in a laboratory.

(*a*) The oxygen inside the cylinder is at a pressure of $2{\cdot}82 \times 10^6$ Pa and a temperature of 19·0 °C.

The cylinder is now moved to a storage room where the temperature is 5·0 °C.

(i) Calculate the pressure of the oxygen inside the cylinder when its temperature is 5·0 °C.

(ii) What effect, if any, does this decrease in temperature have on the density of the oxygen in the cylinder?

Justify your answer.

(*b*) (i) The volume of oxygen inside the cylinder is $0{\cdot}030\,\text{m}^3$.

The density of the oxygen inside the cylinder is $37{\cdot}6\,\text{kg}\,\text{m}^{-3}$.

Calculate the mass of oxygen in the cylinder.

(ii) The valve on the cylinder is opened slightly so that oxygen is gradually released.

The temperature of the oxygen inside the cylinder remains constant.

Explain, in terms of particles, why the pressure of the gas inside the cylinder decreases.

(iii) After a period of time, the pressure of the oxygen inside the cylinder reaches a constant value of $1{\cdot}01 \times 10^5$ Pa. The valve remains open.

Explain why the pressure does not decrease below this value.

Candidate A's answer:

(a) (i) $\frac{P_1}{T_1} = \frac{P_2}{T_2}$

$\Rightarrow \frac{2{\cdot}82 \times 10^6}{19} = \frac{P_2}{5}$

$\Rightarrow P_2 = 7{\cdot}42 \times 10^5$ Pa

(a) (ii) *density* $= \frac{\textit{mass}}{\textit{volume}}$

Neither the mass or volume of the gas have changed (because the cylinder is a sealed, rigid container). The density is therefore the same.

(b) (i) $m = 37{\cdot}6 \times 0{\cdot}03 = 1{\cdot}13$ kg

(b) (ii) The molecules have more room. There are fewer collisions and so the force is smaller.

(b) (iii) This is the value of air pressure.

Candidate B's answer:

(a) (i) $\frac{P_1}{T_1} = \frac{P_2}{T_2}$

$T_1 = 19 + 273 = 292$ K $\quad T_2 = 5 + 273 = 278$ K

$\Rightarrow \frac{2{\cdot}82 \times 10^6}{292} = \frac{P_2}{278}$

$\Rightarrow P_2 = 2{\cdot}68 \times 10^6$ Pa

(a) (ii) Neither the mass or volume of the gas have changed. The density is therefore the same.

(b) (i) $\rho = \frac{m}{V}$

$37{\cdot}6 = \frac{m}{0{\cdot}03}$

$m = 37{\cdot}6 \times 0{\cdot}03 = 1{\cdot}13$ kg

(b) (ii) The molecules are moving slower and colliding less hard with the walls. This causes a smaller force and pressure on the walls.

(b) (iii) The inside pressure is now the same as the outside, atmospheric pressure.

Analysis of example answers

Candidate A's answer:

Part	Strengths	Weaknesses
(a)(i)	• the correct formula has been quoted	• the temperatures have not been changed into Kelvins – this means that most marks are lost
(a)(ii)	• a very good answer, clearly and comprehensively explained	
(b)(i)	• the correct answer	• the formula has not been quoted. It is always good practice to quote the relevant formula because this maximises the marks you can gain if you make an error.
(b)(ii)		• this answer is lacking any detail – there is no mention of the important collisions being *with the walls* of the container or that it is the number of these collisions *per second* which affects the pressure
(b)(iii)		• air can have any value of pressure – the candidate should have used the term 'atmospheric pressure'

Candidate B's answer:

Part	Strengths	Weaknesses
(a)(i)	• a correct answer – clear and complete	
(a)(ii)	• no change in mass and volume is correct • the conclusion is correct	• the explanation is incomplete – the formula for density has been omitted

Continued

(b)(i)	• a correct answer – clear and complete	
(b)(ii)		• wrong Physics – constant temperature means there is no change in the speed of the molecules
(b)(iii)	• the correct answer	

The 'perfect' answer would be:

(a) (i) Answer B, (a) (ii) Answer A, (b) (i) Answer B

(b) (ii) There are fewer molecules in the cylinder. These molecules therefore cause fewer collisions with the walls per second. This means there is a smaller force on the walls and therefore a smaller pressure.

(b) (iii) Answer B.

QUESTION 5

The diagram below shows the basic features of a proton accelerator. It is enclosed in an evacuated container.

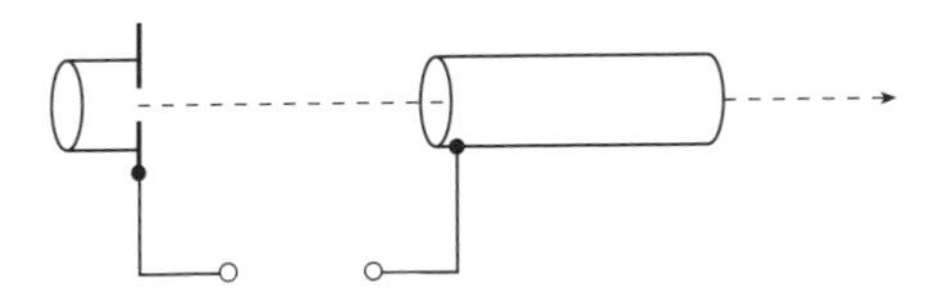

Protons released from the proton source start from rest at **P**.
A potential difference of 200 kV is maintained between **P** and **Q**.

(*a*) What is meant by the term *potential difference of 200 kV*?

(*b*) Explain why protons released at **P** are accelerated towards **Q**.

(*c*) Calculate:

(i) the work done on a proton as it accelerates from **P** to **Q**;

(ii) the speed of a proton as it reaches **Q**.

(*d*) The distance between **P** and **Q** is now halved.

What effect, if any, does this change have on the speed of a proton as it reaches **Q**? Justify your answer.

Candidate A's answer:

(a) *Each coulomb of charge receives 200 000 J of energy as it moves from P to Q.*

(b) *Protons have a positive charge.*

They therefore experience an attractive force towards Q because it is negative.

(c) (i) $W = QV$

$= 1{\cdot}6 \times 10^{-19} \times 200$

$= 3{\cdot}2 \times 10^{-17}\ J$

(c) (ii) $\frac{1}{2}mv^2 = W$

$\Rightarrow \frac{1}{2} \times 1{\cdot}763 \times 10^{-27} \times v^2 = 3{\cdot}2 \times 10^{-17}$

$\Rightarrow v = 1{\cdot}9 \times 10^5\ ms^{-1}$

(d) *There is a bigger force on the protons and so they go faster.*

Candidate B's answer:

(a) *A charge gains 200 000 J of energy when it travels from P to Q.*

(b) *Protons are attracted towards Q.*

(c) (i) $W = QV$

$= 1{\cdot}6 \times 10^{-19} \times 200\,000$

$= 3{\cdot}2 \times 10^{-14}\ J$

(c) (ii) $E_k = \frac{1}{2}mv^2 = W$

$\Rightarrow \frac{1}{2} \times 1{\cdot}673 \times 10^{-27} \times v^2 = 3{\cdot}2 \times 10^{-14}$

$\Rightarrow v^2 = 3{\cdot}83 \times 10^{13}$

$\Rightarrow v = 6{\cdot}2 \times 10^6\ ms^{-1}$

(d) *There is no effect on the speed of a proton.*

The kinetic energy is found from $Q \times V$*, as above. Neither Q nor V have changed, so the* E_k *and the speed remain the same.*

Analysis of example answers

Candidate A's answer:

Part	Strengths	Weaknesses
(a)	• the correct answer	
(b)	• the correct answer	
(c)(i)	• the correct formula has been chosen	• the prefix 'kilo' has been missed
(c)(ii)	• the correct relationship has been used	• the figures '6' and '7' have been transposed in the value for the mass of a proton – this is a wrong substitution and is a costly error
(d)	• it is true that the force is greater	• the conclusion is wrong because the final speed depends on more than the force. (As the force acts for a shorter time, the increase in speed is the same as before)

Candidate B's answer:

Part	Strengths	Weaknesses
(a)		• the answer must say *one coulomb of charge*
(b)	• a true statement BUT	• no mention of proton's positive charge • no reason is given to explain *why* the protons are attracted to Q
(c)(i)	• the correct answer	
(c)(ii)	• the correct answer	
(d)	• the correct answer and explanation	

The 'perfect' answer would be:

(a) Answer A, (b) Answer A, (c) (i) Answer B, (c) (ii) Answer B,
(c) (iii) Answer B.

QUESTION 6

26. A 12 volt battery of negligible internal resistance is connected in a circuit as shown.

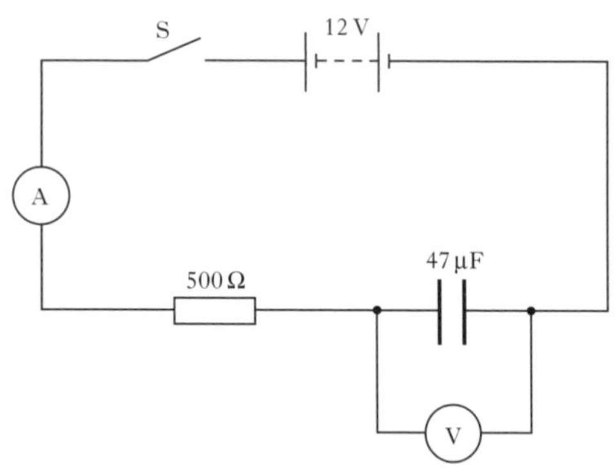

The capacitor is initially uncharged. Switch S is then closed and the capacitor starts to charge.

(*a*) Sketch a graph of the current against time from the instant switch S is closed. Numerical values are not required.

(*b*) At one instant during the charging of the capacitor the reading on the ammeter is 5·0 mA.

Calculate the reading on the voltmeter at this instant.

(*c*) Calculate the **maximum** energy stored in the capacitor in this circuit.

(*d*) The 500 Ω resistor is now replaced with a 2·0 kΩ resistor.

What effect, if any, does this have on the maximum energy stored in the capacitor?

Justify your answer.

Candidate A's answer:

(a)

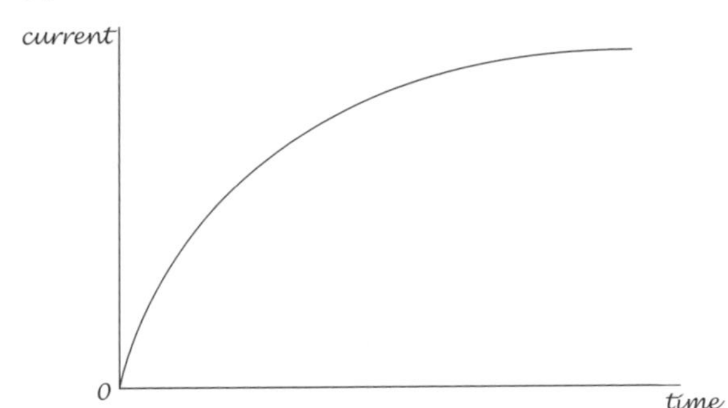

(b) $V = IR = 5.0 \times 10^{-3} \times 500 = 2.5\ V$

(c) $E = \frac{1}{2}CV^2$

$= 0.5 \times 47 \times 10^{-6} \times 2.5^2$

$= 1.47 \times 10^{-4}\ J$

(d) A larger resistance means the current is smaller. This means the energy is less.

Candidate B's answer:

(a)

current

0

time

(b) $V_R = IR = 5.0 \times 10^{-3} \times 500 = 2.5$

Voltmeter reading $= V_C = 12 - 2.5 = 9.5\ V$

(c) $E = \frac{1}{2}CV^2$

$= 0.5 \times 47 \times 10^{-6} \times 12^2$

$= 3.4 \times 10^{-3}\ J$

(d) $E = \frac{1}{2}CV^2$

Neither C or V have changed, so the energy stored is the same as before.

Analysis of example answers

Candidate A's answer:

Part	Strengths	Weaknesses
(a)	• the axes and origin have been labelled	• the line wrongly curves *upwards* – this is what happens to the voltage across the plates of the capacitor as it charges
(b)	• this calculation is a necessary part of finding the answer	• the answer is incomplete – this is the voltage across the resistor, but the voltmeter reads the voltage across the capacitor
(c)	• a correct formula has been quoted	• the wrong value of voltage has been used – *maximum* energy is stored when the voltage across the capacitor is equal to the p.d. across the power supply
(d)		• this is the wrong conclusion – the value of current is irrelevant to the quantity of energy stored

Candidate B's answer:

Part	Strengths	Weaknesses
(a)	• the correct answer	
(b)	• this is the correct method and answer	
(c)	• this is the correct answer	
(d)	• the correct answer, clearly explained by referring to the formula and saying what happens to each of the quantities	

The 'perfect' answer would be: Answer B.

QUESTION 7

(*a*) A microphone is connected to the input terminals of an oscilloscope.
A tuning fork is made to vibrate and held close to the microphone as shown.

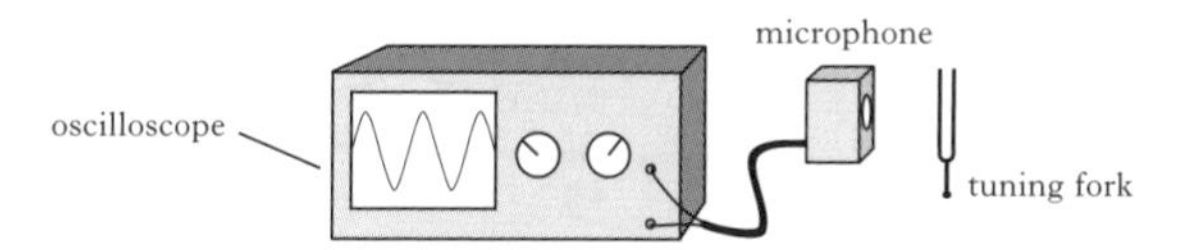

The following diagram shows the trace obtained and the settings on the oscilloscope.

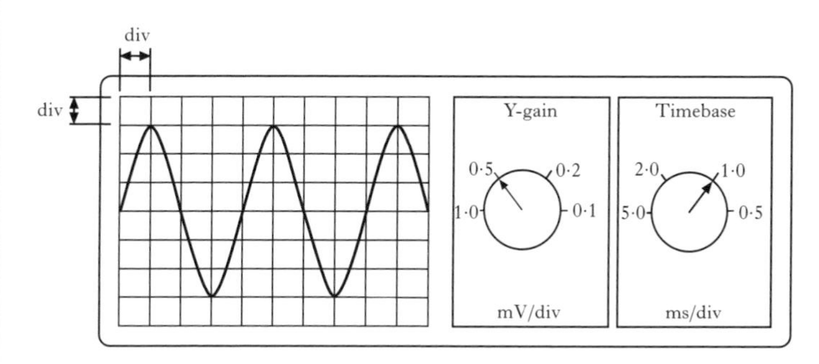

Calculate:

(i) the peak voltage of the signal;

(ii) the frequency of the signal.

(*b*) To amplify the signal from the microphone, it is connected to an op-amp circuit. The oscilloscope is now connected to the output of the amplifier as shown.

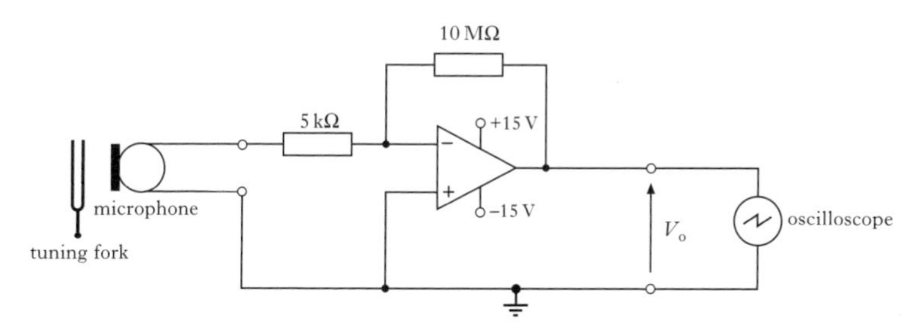

The settings of the oscilloscope are adjusted to show a trace of the amplified signal.

(i) In which mode is this op-amp being used?

(ii) The peak voltage from the microphone is now 6·2 mV.

Calculate the **r.m.s.** value of the output voltage, V_o, of the op-amp.

(iii) With the same input signal and settings on the oscilloscope, the supply voltage to the op-amp is now reduced from ± 15 V to ± 9 V.

What effect does this change have on the trace on the oscilloscope?

Justify your answer.

Candidate A's answer:

(a) (i) $V_p = 6 \times 0{\cdot}5 = 3{\cdot}0\ mV$

(ii) $T = 2 \times 1{\cdot}0 \times 10^{-3}$

$= 2 \times 10^{-3}$

$f = \frac{1}{T}$

$= \frac{1}{2 \times 10^{-3}} = 125\ Hz$

(b) (i) inverting

(ii) peak $V_{out} = -6{\cdot}2 \times 10^{-3} \times \frac{10 \times 10^6}{5 \times 10^3}$

$= -12{\cdot}4\ V$

$V_{rms} = \frac{V_p}{\sqrt{2}}$

$= \frac{12{\cdot}4}{\sqrt{2}} = 8{\cdot}8\ V$

(iii) The output voltage becomes a square wave. This is because the peak output voltage saturates at about 9.0 volts.

Candidate B's answer:

(a) (i) $V_p = 3 \times 0{\cdot}5 = 1{\cdot}5\ mV$

(ii) period $= 4 \times 1{\cdot}0 \times 10^{-3}$

$= 4{\cdot}0 \times 10^{-3}$

frequency $= \frac{1}{\text{period}}$

$= \frac{1}{4{\cdot}0 \times 10^{-3}} = 250\ Hz$

(b) (i) inverting

(ii) peak $V_{out} = -V_{in} \times \frac{10 \times 10^6}{5 \times 10^3}$

$= -6{\cdot}2 \times 10^{-3} \times \frac{10 \times 10^6}{5 \times 10^3}$

$= -12{\cdot}4\ V$

$V_{rms} = \frac{Vp}{\sqrt{2}}$

$= \frac{-12{\cdot}4}{\sqrt{2}} = -8{\cdot}8\ V$

(iii) The tops and bottoms of the trace are now flattened because the op-amp saturates when the peak output voltage is ±9.0 V

Analysis of example answers

Candidate A's answer:

Part	Strengths	Weaknesses
(a)(i)		• the total 'height' of the wave has been used – this is incorrect, it should be half the 'height'
(a)(ii)		• only half the 'wavelength' of the wave has been used – this is incorrect, it should be 4 'divs' for this wave
(b)(i)	• the correct answer	
(b)(ii)	• the correct answer	• no formula has been given for calculating peak V_{out}. This would have been costly if the numbers had been incorrect.
(b)(iii)		• although the output voltage is 'flattened' or 'clipped' it is *not square* because low input voltages are not amplified enough • it is wrong to say 'the output voltage saturates' – it is the *op–amp* which saturates

Candidate B's answer:

Part	Strengths	Weaknesses
(a)(i)	• the correct answer	
(a)(ii)	• the correct answer	
(b)(i)	• the correct answer	
(b)(ii)	• the correct answer	
(b)(iii)	• a correct answer and explanation	

The 'perfect' answer would be: Answer B.

QUESTION 8

27. A laser produces a narrow beam of monochromatic light.

(*a*) Red light from a laser passes through a grating as shown.

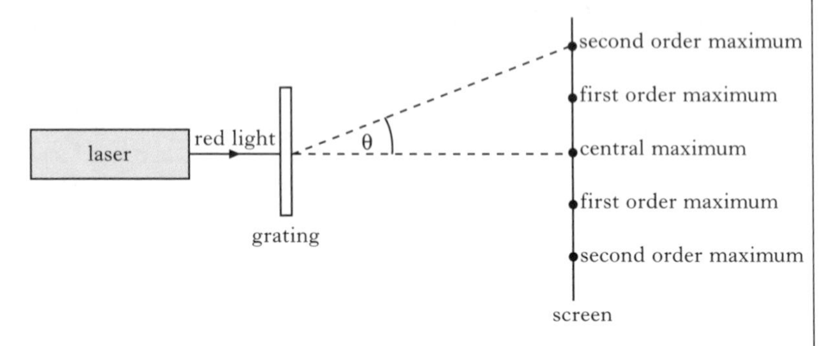

A series of maxima and minima is observed.

Explain in terms of waves how a **minimum** is produced.

(*b*) The laser is now replaced by a second laser, which emits blue light.

Explain why the observed maxima are now closer together.

Explain why the observed maxima are now closer together.

(*c*) The wavelength of the blue light from the second laser is $4{\cdot}73 \times 10^{-7}$ m. The spacing between the lines on the grating is $2{\cdot}00 \times 10^{-6}$ m.

Calculate the angle between the central maximum and the second order maximum.

Candidate A's answer:

(a) The waves are not in phase when they meet.

(b) Blue light diffracts less than red light.

(c) $n\lambda = d \sin\theta$

$\Rightarrow 2 \times 4{\cdot}73 \times 10^{-7} = 2{\cdot}00 \times 10^{-6} \times \sin\theta$

$\Rightarrow \sin\theta = 2 \times \dfrac{4{\cdot}73 \times 10^{-7}}{2{\cdot}00 \times 10^{-6}} = 0{\cdot}473$

So, $\theta = 28{\cdot}2°$

Candidate B's answer:

(a) *The waves are 180° out of phase and so destructive interference occurs.*

(b) *Blue light has a shorter wavelength than red light.*

(c) $n\lambda = d \sin\theta$

$\Rightarrow 1 \times 4{\cdot}73 \times 10^{-7} = 2{\cdot}00 \times 10^{-6} \times \sin\theta$

$\sin\theta = \dfrac{4{\cdot}73 \times 10^{-7}}{2{\cdot}00 \times 10^{-6}} = 0{\cdot}2365$

So, $\theta = 13{\cdot}7°$

Analysis of example answers

Candidate A's answer:

Part	Strengths	Weaknesses
(a)		• 'not in phase' does not necessarily mean that destructive interference occurs For example, [diagram of two sine waves slightly offset] these two waves are 'not in phase', but they do not cancel each other out
(b)		• For constructive interference, path difference must $= n\lambda$. The shorter wavelength blue light therefore interferes constructively closer to the straight through direction than the red (Note that *both* colours diffract into semicircular waves on the other side of the grating)
(c)	• the correct answer	

Candidate B's answer:

Part	Strengths	Weaknesses
(a)	• the correct answer – it is good to make clear that the waves are completely out of phase by using a description such as '180° out of phase' OR 'perfectly out of phase' OR 'half a wavelength out of phase'	
(b)	• the correct answer	
(c)	• the correct formula has been chosen	• '*n*' has been wrongly substituted as '1' – the *second* order maximum means that '*n*' should be '2' (Note – in this formula, '*n*' stands for the order of the spectrum. Do not confuse this with the formulas in which '*n*' is used for refractive index)

The 'perfect' answer would be: (a) Answer B, (b) Answer B, (c) Answer A.

QUESTION 9

A ray of red light is directed at a glass prism of side 80 mm as shown in the diagram below.

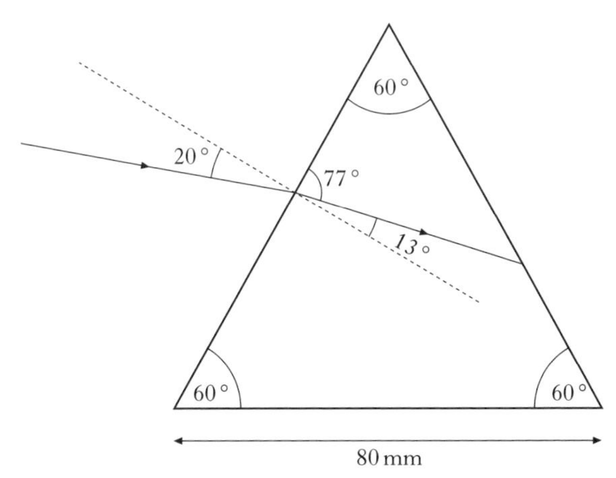

(*a*) Using information from this diagram, show that the refractive index of the glass for this red light is 1·52.

(*b*) What is meant by the term *critical angle*?

(*c*) Calculate the critical angle for the red light in the prism.

(*d*) Sketch a diagram showing the path of the ray of red light until after it leaves the prism. Mark on your diagram the values of all relevant angles.

Candidate A's answer:

(a) $n = \frac{\sin \theta_1}{\sin \theta_2} = \frac{\sin 20}{\sin 13} = 1{\cdot}52$

(b) The critical angle is the angle of incidence when total internal reflection occurs.

(c) $\theta_c = \frac{1}{n} = \frac{1}{1{\cdot}52} = 41°$

(d) Angle at right hand face = 43°
Angle of incidence at right hand face = 47°
This is greater than the critical angle so total internal reflection occurs at this face.

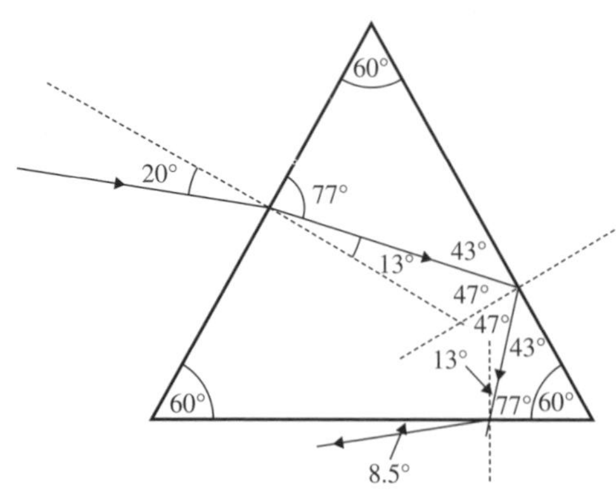

Angle between ray and bottom face = 77°
Angle of incidence at bottom face = 13°

Refraction at bottom face:
$n = \sin\theta_1 / \sin\theta_2$
$1.52 = \sin 13 / \sin\theta_2$
$\theta_2 = 8.5°$

Candidate B's answer:

(a) $n_1 \sin\theta_1 = n_2 \sin\theta_2$
$1{\cdot}0 \times \sin 20 = n_2 \sin 13$
$n_2 = 1{\cdot}0 \times \frac{\sin 20}{\sin 13} = 1{\cdot}52$

(b) *The critical angle is the value of the angle of incidence when the angle of refraction is 90°.*

(c) $\sin\theta_c = \frac{1}{n} = \frac{1}{1{\cdot}52} = 0{\cdot}657894736$ *giving* $\theta_c = 41°$

(d) *Angle between ray and right hand face* $= 180 - (60 + 77) = 43°$
Angle of incidence at right hand face $= 90 - 43 = 47°$
This is greater than the critical angle so total internal reflection occurs at this face.

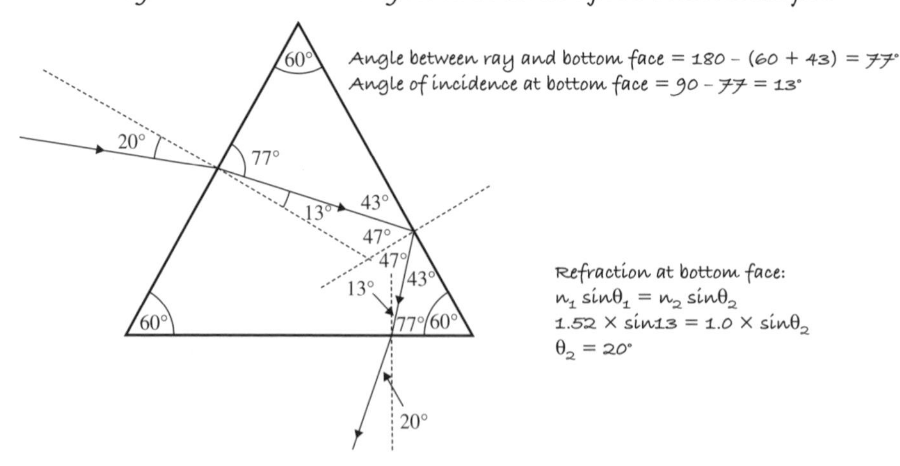

Analysis of example answers

Candidate A's answer:

Part	Strengths	Weaknesses
(a)	• a correct solution	
(b)		• an incorrect statement – (total internal reflection does not occur until the angle of incidence is *greater* than the critical angle)
(c)		• the initial relationship is wrong – it should be $\sin \theta_c = \frac{1}{n}$
(d)	• the angles of incidence at both the right hand face and the bottom face have been worked out correctly	• the refraction formula has been applied incorrectly – the candidate has not allowed for the fact that the ray is now travelling *out* of the glass into air • the angle of refraction has been shown on the diagram as being between the ray and the surface of the glass – it should be between the ray and the normal

Candidate B's answer:

Part	Strengths	Weaknesses
(a)	• a correct solution	
(b)	• a correct definition of critical angle	
(c)	• the correct answer	
(d)	• a clear and correctly explained analysis • the use of $n_1 \sin \theta_1 = n_2 \sin \theta_2$ has reduced the chances that the candidate gets mixed up about which angle 'goes with' which refractive index	

Candidate B has given a 'perfect' answer.

QUESTION 10

29. Ultraviolet radiation from a lamp is incident on the surface of a metal.

This causes the release of electrons from the surface of the metal.

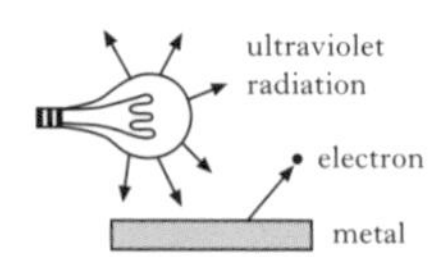

The energy of each photon of ultraviolet light is $5{\cdot}23 \times 10^{-19}$ J.

The work function of the metal is $2{\cdot}56 \times 10^{-19}$ J.

(*a*) Calculate:

(i) the maximum kinetic energy of an electron released from this metal by this radiation;

(ii) the maximum speed of an emitted electron.

(*b*) The source of ultraviolet radiation is now moved further away from the surface of the metal.

State the effect, if any, this has on the maximum speed of an emitted electron.

Justify your answer.

Candidate A's answer:

(a) (i) *kinetic energy of electron* $= E_{photon}$
$= 5{\cdot}23 \times 10^{-19}$ J

(ii) $E_k = \frac{1}{2}mv^2$

$\frac{1}{2}mv^2 = 5{\cdot}23 \times 10^{-19}$

$\frac{1}{2} \times 9{\cdot}11 \times 10^{-31} \times v^2 = 5{\cdot}23 \times 10^{-19}$

$v^2 = 2 \times \frac{5{\cdot}23 \times 10^{-19}}{9{\cdot}11 \times 10^{-31}}$

$v = 1071536\ \text{m s}^{-1}$

(b) *The maximum speed will be slower because less energy reaches the surface of the metal.*

Candidate B's answer:

(a) (i) $E_k = E_{photon}$ − *work function*
$= 5{\cdot}23 \times 10^{-19} - 2{\cdot}56 \times 10^{-19}$
$= 2{\cdot}67 \times 10^{-19}$ J

(ii) $E_k = \frac{1}{2}mv^2$

$\frac{1}{2}mv^2 = 2{\cdot}67 \times 10^{-19}$

$\frac{1}{2} \times 9{\cdot}11 \times 10^{-31} \times v^2 = 2{\cdot}67 \times 10^{-19}$

$v^2 = 2 \times \frac{2{\cdot}67 \times 10^{-19}}{9{\cdot}11 \times 10^{-31}}$

$v = 7{\cdot}66 \times 10^5\ \text{m s}^{-1}$

(b) *Each photon still has the same energy as before. Each electron therefore receives the same energy as before and so the maximum speed is the same as before.*

Analysis of example answers

Candidate A's answer:

Part	Strengths	Weaknesses
(a)(i)		• the *kinetic* energy of an electron is not equal to the energy of a photon – (the *total* energy received by an electron is equal to the energy of a photon)
(a)(ii)	• the correct formula has been chosen • the answer from part (a)(i) has been used as the value of energy • the mass of an electron has been correctly taken from the data sheet on page two of the examination paper	• the answer has been quoted to too many significant figures (seven) – (as the data is given to three sig. figs., so should the answer)
(b)		• the conclusion is wrong – it is true that the *total* energy on the metal's surface is less. However, there is still one photon colliding with one electron, meaning that any electron receives the same energy as before

Candidate B's answer:

Part	Strengths	Weaknesses
(a)(i)	• the correct answer	
(a)(ii)	• the correct answer	
(b)	• a correct answer, clearly explained	

The 'perfect' answer would be: Answer B.

QUESTION 11

A fission reaction is shown by the following statement.

$$^{235}_{92}U + ^{1}_{0}n \longrightarrow ^{140}_{58}Ce + ^{r}_{40}Zr + 2\,^{1}_{0}n + s\,^{0}_{-1}e$$

(i) Give the values of the numbers represented by the letters ***r*** and ***s*** in the above statement. **1**

(ii) Explain why energy is released by a fission reaction. **1**

(iii) The masses of the particles in the above reaction are shown in the table.

Particle	*Mass/kg*
$^{235}_{92}U$	390.173×10^{-27}
$^{140}_{58}Ce$	232.242×10^{-27}
$^{r}_{40}Zr$	155.884×10^{-27}
$^{1}_{0}n$	1.675×10^{-27}
$^{0}_{-1}e$	negligible

Calculate the energy released in this reaction. **3**

Candidate A's answer:

(a) *(i)* $r = (235 + 1) - (140 + 2) = 94$

$s = -\{92 - (58 + 40)\} = -\{-6\} = 6$

(a) *(ii) The total mass of the nuclei after the reaction is less than the total mass of the nuclei before the reaction. The loss in mass is converted into energy according to Einstein's equation,* $E = mc^2$.

(a) *(iii) Total mass before* $= 390.173 \times 10^{-27} + 1.675 \times 10^{-27} = 3.91848 \times 10^{-25}$ *(kg)*

Total mass after $= 232.242 \times 10^{-27} + 155.884 + 10^{-27} + (2 \times 1.675 \times 10^{-27})$

$= 3.91476 \times 10^{-25}$ *(kg)*

$\Delta m = 3.91848 \times 10^{-25} - 3.91476 \times 10^{-25} = 3.72 \times 10^{-28}$ *(kg)*

$E = mc^2 = 3.72 \times 10^{-28} \times (3 \times 10^8)^2 = 3.348 \times 10^{-11}$ *J*

Candidate B's answer:

(a) (i) $r = 94 \quad s = 6$

(a) (ii) Energy is released according to $E = mc^2$*, a formula produced by Einstein.*

(a) (iii) mass before $= 390.173 \times 10^{-27} + 1.675 \times 10^{-27} = 3.91848 \times 10^{-25}$
$= 3.918 \times 10^{-25}$

mass after $= 232.242 \times 10^{-27} + 155.884 \times 10^{-27}$
$+ (2 \times 1.675 \times 10^{-27})$
$= 3.91476 \times 10^{-25} = 3.915 \times 10^{-25}$

$\Delta m = 3.918 \times 10^{-25} - 3.915 \times 10^{-25} = 0.003 \times 10^{-25}$

$E = mc^2 = 0.003 \times 10^{-25} \times (3 \times 10^8)^2 = 2.7 \times 10^{-11}$ J

Analysis of example answers

Candidate A's answer:

Part	Strengths	Weaknesses
(a)(i)	• the correct answer • a full 'explanation' is given (although not required for one mark)	
(a)(ii)	• a full and correct explanation	
(a)(iii)	• a full and detailed analysis • working is very clearly set out, each step being shown (*if* an arithmetical mistake had been made, a marker would have been able to easily identify the fact and minimum marks would have been lost)	

Candidate B's answer:

Part	Strengths	Weaknesses
(a)(i)	• the correct answers	• no working shown – no chance of gaining partial marks (if available) if an error had been made
(a)(ii)	• an appropriate reference to $E = mc^2$	• it has not been made clear that there is a loss in mass during the reaction
(a)(iii)	• an appropriate attempt to use $E = mc^2$	• the total mass both before and after the reaction have been rounded – because the change in mass is so small, this is introduces very large inaccuracies - this is wrong Physics and receives the high penalty of losing most of the marks

The 'perfect' answer would be: Answer A (for all parts).

8 Appendix A – Practice at Calculations

Every year the majority of candidates in the national examination perform well in most of the questions which require the use of formulas to calculate numerical answers.

You need to ensure that you, too, are good at this task. The following questions are designed to give you extensive practice at using the formulas met during study of the Higher Physics course. While answering them, remember to use the technique (whether written down or done in your head) of making a list of the given quantities plus the unknown – then use this list to identify which formula to apply.

For loads more practice calculation questions and answers, go to www.leckieandleckie.co.uk/7248calc.pdf

QUESTIONS

1. Calculate the speed (in $m\,s^{-1}$) of a vehicle which travels 5·3 km in a time of 30 minutes.
2. An object speeds up from 3·0 $m\,s^{-1}$ to 45 $m\,s^{-1}$ in a time of 4·3 s. Calculate its acceleration.
3. An object has an initial speed of 12·0 $m\,s^{-1}$. It then accelerates at 5·30 $m\,s^{-2}$ for a time of 81·2 s. How fast is it travelling at the end of this time?
4. An object is dropped near the earth's surface. How fast is the object falling 5·6 s later? Ignore the effects of air resistance.
5. An object is dropped near the earth's surface. How far will the object have fallen 8·2 s later? Ignore the effects of air resistance.
6. An object is dropped from a height of 78 m above the earth. How long does it take for the object to reach the ground? Ignore the effects of air resistance.
7. A car takes 6·8 s to speed up uniformly from 2·5 $m\,s^{-1}$ to 23·4 $m\,s^{-1}$. How far does the car travel in this time?
8. An object travels a distance of 36 m while it speeds up from 1·5 $m\,s^{-1}$ to 6·9 $m\,s^{-1}$. Calculate its acceleration.

9. Calculate the momentum of a ship of mass $4{\cdot}5 \times 10^5$ kg when it is travelling at 2·3 m s^{-1}.

10. An object of mass 5·5 kg has a momentum of 74·0 kg m s^{-1}. What is its speed?

ANSWERS

Q	Answer
1	$\text{speed} = \frac{\text{distance}}{\text{time}} = \frac{5{,}300}{(30 \times 60)} = 2{\cdot}9 \text{ m s}^{-1}$
2	$\text{acceleration} = \frac{\text{change in speed}}{\text{time}} = \frac{42}{4{\cdot}3} = 9{\cdot}76744 = 9{\cdot}8 \text{ m s}^{-2}$
3	$v = u + at = 12{\cdot}0 + (5{\cdot}30 \times 81{\cdot}2) = 12{\cdot}0 + 430{\cdot}36 = 442 \text{ m s}^{-1}$
4	$v = u + at = 0 + (9{\cdot}8 \times 5{\cdot}6) = 54{\cdot}88 = 55 \text{ m s}^{-1}$
5	$s = ut + \frac{1}{2}at^2 = 0 + 0{\cdot}5 \times 9{\cdot}8 \times (8{\cdot}2)^2 = 329{\cdot}476 = 330 \text{ m}$
6	$s = ut + \frac{1}{2}at^2$ $\Rightarrow 78 = 0 + 0{\cdot}5 \times 9{\cdot}8 \times (t)^2 \Rightarrow t^2 = \frac{(2 \times 78)}{9{\cdot}8} \Rightarrow t = 4{\cdot}0 \text{ s}$
7	$s = \frac{(u + v)}{2} t = \frac{(2{\cdot}5 + 23{\cdot}4)}{2} \times 6{\cdot}8 = 88{\cdot}06 = 88{\cdot}1 \text{ m}$
8	$v^2 = u^2 + 2as \Rightarrow a = \frac{v^2 - u^2}{2s} = \frac{(6{\cdot}9)^2 - (1{\cdot}5)^2}{2 \times 36} = 0{\cdot}63 \text{ m s}^{-2}$
9	$\text{momentum} = \text{mass} \times \text{velocity} = 4{\cdot}5 \times 10^5 \times 2{\cdot}3 = 1{\cdot}035 \times 10^6$ $= 1{\cdot}04 \times 10^6 \text{ kg m s}^{-1}$
10	$\text{velocity} = \frac{\text{momentum}}{\text{mass}} = \frac{74{\cdot}0}{5{\cdot}5} = 13{\cdot}454545 = 13{\cdot}5 \text{ m s}^{-1}$

9 Appendix B – Practice at Problem Solving

Your problem solving skills will be improved by frequent and repeated practice at answering questions which take you out of your 'comfort zone'. This means that you need to attempt to answer questions which you find 'hard' (at least, when you first try them). These are questions which, when you start to answer them, you cannot 'see' all the way to the final answer. In these circumstances, you need to think through the content of the Higher Physics course, trying to identify which part(s) is/are relevant.

The following questions are designed to help you develop this skill of identifying the relevant Physics for a given situation.

For loads more practice at problem solving questions and answers, go to www.leckieandleckie.co.uk/7248prob.pdf

QUESTIONS

1. Why is there little point in buying a hand-operated stopwatch which can read to 0·01 s?
2. An athlete completes a circuit of a running track. Explain why the average velocity of the athlete is zero.
3. A rocket takes off vertically from the surface of the earth. Explain why its acceleration increases during the first few seconds after launch. (Give at least two reasons.)
4. Why is a displacement-time graph curved in shape while an object is accelerating?
5. Explain why air resistance reduces the horizontal range of a projectile.
6. The acceleration due to gravity is 9·8 $m\,s^{-2}$ for all objects near the surface of the earth. Why, then, does a feather take longer to fall than a hammer from the same height?
7. Why is the path of a golf ball not the parabolic path predicted by basic Physics?
8. A spacecraft, far from the earth, is drifting without its motors on. It is found to be accelerating straight ahead. What is the likely reason for this?
9. Total momentum is conserved in all collisions. What has happened to the momentum of a lorry which has stopped after crashing into a tree?

10. Two protective helmets are being tested. 'A' has little padding. 'B' is thickly padded. They are both subjected to a crash test at the same speed. Using 'impulse', explain which helmet protects a person's head better.

ANSWERS

1. Human reaction time can introduce an uncertainty of around 0·2 or 0·3 s. This is much larger than the hundredth of a second that the electronics can measure to and so the second decimal place is meaningless when it is hand operated.

2. Completion of the lap means that (s)he is back where (s)he started. Thus the displacement is zero. Since average velocity $= \frac{\text{displacement}}{\text{time}}$, zero displacement means zero velocity.

3. $a = \frac{F}{m}$, m decreases due to fuel being used up
   F increases since the downward weight decreases (as m decreases)

   Smaller effects are also that air resistance decreases as the atmosphere becomes thinner and at greater heights the gravitational field strength is smaller.

4. An accelerating object is one whose speed is changing. It will therefore move different distances each second. These varying increases in distance for equal steps in time cause the graph to curve.

5. Air resistance is a frictional force which acts against motion. Horizontal speed is therefore reduced and so the horizontal distance travelled (the range) also reduces.

6. $a = 9{\cdot}8\ \text{m s}^{-2}$ for all objects only when the effects of air resistance are ignored. A feather, having a small weight and large surface area, soon experiences a large air resistance upwards acting against its weight. The feather therefore tends to drift down with a low value of terminal velocity.

7. air resistance is not negligible – this changes the golf ball's horizontal and vertical speeds from the ideal values

   the spin of the ball introduces other effects (beyond the scope of Higher Physics) which affect its path

8. There is likely to be a planet straight ahead. The gravitational field strength of the planet causes an attractive force, making the craft accelerate towards it.

9. The momentum of the lorry has been shared with the tree and therefore with the whole of the earth. In fact, the rotation of the earth will have been (very slightly) affected by the collision.

10. Helmet 'B' is better. Crashes from the same speed mean that the change in momentum is the same for both. Therefore the impulse is the same for both. But the thick padding increases the time of impact and so decreases the average force on the head because impulse = change in momentum.